RESILIENCE IN THE ANTHROPOCENE

This book offers the first critical, multi-disciplinary study of how the concepts of resilience and the Anthropocene have combined to shape contemporary thought and governmental practice.

Faced with the climate catastrophe of the Anthropocene, theorists and policymakers are increasingly turning to 'sustainable', 'creative' and 'bottom-up' imaginaries of governance. The book brings together cutting-edge insights from leading geographers, international relations scholars and philosophers to explore how the concepts of resilience and the Anthropocene challenge and transform prevailing understandings of Earth, space, time and knowledge, and how these transformations reshape governance, ethics and critique today. This book examines how the Anthropocene calls into question established categories through which modern societies have tended to make sense of the world and engage in critical reflection and analysis. It also considers how resilience approaches attempt to re-stabilise these categories – and the ethical and political effects that result from these resilience-based efforts.

Offering innovative insights into the problem of how environmental change is known and governed in the Anthropocene, this book will be of interest to students in fields such as geography, international relations, anthropology, science and technology studies, sociology and the environmental humanities.

David Chandler is Professor of International Relations, University of Westminster, UK. His recent monographs include *Becoming Indigenous: Governing Imaginaries in the Anthropocene* (with Julian Reid, 2019) and *Ontopolitics in the Anthropocene: An Introduction to Mapping, Sensing and Hacking* (2018).

Kevin Grove is Associate Professor of Geography at Florida International University, USA. His research explores the politics of disaster management and resilience in the Caribbean and North American cities. He is the author most recently of *Resilience* (Routledge Key Ideas in Geography series, 2018).

Stephanie Wakefield is an Urban Studies Foundation International Postdoctoral Fellow based at Florida International University, USA. Her work explores experimental practices for living in and governing the Anthropocene. Her book *Anthropocene Back Loop: Experimentation in Unsafe Operating Space* is forthcoming.

ROUTLEDGE RESEARCH IN THE ANTHROPOCENE

Series Editors: Jamie Lorimer and Kathryn Yusoff

The Routledge Research in the Anthropocene Series offers the first forum for original and innovative research on the epoch and events of the Anthropocene. Titles within the series are empirically and/or theoretically informed and explore a range of dynamic, captivating and highly relevant topics, drawing across the humanities and social sciences in an avowedly interdisciplinary perspective. This series will encourage new theoretical perspectives and highlight ground-breaking inter-disciplinary research that reflects the dynamism and vibrancy of current work in this field. The series is aimed at upper-level undergraduates, researchers and research students as well as academics and policy-makers.

Hope and Grief in the Anthropocene
Re-conceptualising Human–Nature Relations
Lesley Head

Releasing the Commons
Edited by Ash Amin and Philip Howell

Climate Change Ethics and the Non-Human World
Edited by Brian G. Henning and Zack Walsh

Involving Anthroponomy in the Anthropocene
On Decoloniality
Jeremy Bendik-Keymer

Resilience in the Anthropocene
Governance and Politics at the End of the World
Edited by David Chandler, Kevin Grove and Stephanie Wakefield

For more information about this series, please visit www.routledge.com/Routle-dge-Research-in-the-Anthropocene/book-series/RRA01

RESILIENCE IN THE ANTHROPOCENE

Governance and Politics at the End of the World

Edited by David Chandler, Kevin Grove and Stephanie Wakefield

Routledge
Taylor & Francis Group

LONDON AND NEW YORK

First published 2020
by Routledge
2 Park Square, Milton Park, Abingdon, Oxon OX14 4RN

and by Routledge
52 Vanderbilt Avenue, New York, NY 10017

Routledge is an imprint of the Taylor & Francis Group, an informa business

British Library Cataloguing-in-Publication Data
A catalogue record for this book is available from the British Library

Library of Congress Cataloging-in-Publication Data
A catalog record has been requested for this book

ISBN: 978-1-138-38742-3 (hbk)
ISBN: 978-1-138-38744-7 (pbk)
ISBN: 978-1-003-03337-0 (ebk)

Typeset in Bembo
by Taylor & Francis Books

CONTENTS

FIGURES

CONTRIBUTORS

Allain Barnett is a post-doctoral associate with the Institute for Water and Environment, Sea Level Solutions Center and the School of International and Public Affairs at Florida International University. His diverse research experience includes aquatic community ecology, ethnographic research in fishing communities, institutional analysis and participatory mapping. His current research interests focus on the intersections between justice, vulnerability and resilience, and the role of infrastructure and institutions in resilience-building initiatives in Miami-Dade County.

Harold Bellanger Rodríguez is a PhD candidate in the Department of Geography at the Université de Montréal. His research is concerned with the politics of race involved in the management of drought in Guatemala.

David Chandler is Professor of International Relations at the University of Westminster. His recent books include *Ontopolitics in the Anthropocene: An Introduction to Mapping, Sensing and Hacking* (Routledge, 2018), *The Neoliberal Subject: Resilience, Adaptation and Vulnerability* (with Julian Reid) (Rowman & Littlefield, 2016) and *Resilience: The Governance of Complexity* (Routledge, 2014).

Nigel Clark is Chair of Social Sustainability at Lancaster University, UK. He is the author of *Inhuman Nature* (Sage, 2011), co-editor (with Kathryn Yusoff) of a *Theory, Culture & Society* special issue on 'Geosocial Formations and the Anthropocene' (2017), and is currently working on a book about *Planetary Social Thought* (with Bronislaw Szerszynski).

Claire Colebrook is Edwin Erle Sparks Professor of English, Philosophy and Women's and Gender Studies at Penn State University. She has written books and articles on contemporary European philosophy, literary history, gender studies, queer theory, visual culture and feminist philosophy. Her most recent book is *Twilight of the Anthropocene Idols* (co-authored with Tom Cohen and J. Hillis Miller) (Open Humanities Press, 2016).

Simon Dalby is Professor of Geography and Environmental Studies at Wilfrid Laurier University, Waterloo, Ontario, where he teaches in the Balsillie School of International Affairs. He is co-editor of *Achieving the Sustainable Development Goals* (Routledge, 2019) and author of *Anthropocene Geopolitics* (University of Ottawa Press, 2020).

Madeleine Fagan is Associate Professor in the Department of Politics and International Studies at the University of Warwick. Her current research explores the politics and ethics of the Anthropocene. She is author of *Ethics and Politics after Poststructuralism: Levinas, Derrida, Nancy* (Edinburgh University Press, 2013/ 2016).

Kevin Grove is Associate Professor of Geography, Department of Global and Sociocultural Studies, Florida International University. His research explores the biopolitics of disaster resilience in Kingston, Jamaica, New York City and Miami. He is the author of *Resilience* (Routledge, 2018) and a number of journal articles published in outlets such as *Progress in Human Geography*, the *Annals of the American Association of Geographers*, *Security Dialogue* and *Environment and Planning D: Society and Space*.

Xochilt Hernández Leiva is an anthropologist who is interested in the relationships between education and hazard management in the global South. She currently teaches at Universidad Americana in Managua, Nicaragua.

Sara Nelson is Killam Postdoctoral Fellow in Geography at the University of British Columbia and a Fellow with the Intergovernmental Platform on Biodiversity and Ecosystem Services. Her research focuses on the political economy of environmental conservation and the history and politics of environmental valuation.

Sébastien Nobert is an Assistant Professor of Human Geography at the Université de Montréal, where he is working on the social, cultural and political dimensions of hazards and risks both in Europe and in the global South. His most recent project looks at the limits of global risk instruments in the management of climate forecasting.

Lauren Rickards is an Associate Professor in the Centre for Urban Research and School of Global, Urban and Social Studies at RMIT University, Melbourne. A Lead Author with the Intergovernmental Panel on Climate Change's forthcoming Sixth Assessment Report, Lauren's cross-disciplinary research examines the sociopolitical and cultural dimensions of climate change and the Anthropocene.

Stephanie Wakefield is an urban geographer and Urban Studies Foundation Research Fellow at Florida International University. Her book *Anthropocene Back Loop: Experimentation in Unsafe Operating Space* is forthcoming from Open Humanities Press. She is working now on a second book project titled *Miami Forever? Urbanism in the Back Loop*, investigating experimental governance practices for living with water in Miami, Florida and through this the emergence of a new paradigm of 'back loop urbanism.'

1

INTRODUCTION

The power of life

Stephanie Wakefield, Kevin Grove and David Chandler

Framing the problem

This book originated out of our collective concern with prevailing treatments of the concepts of resilience and the Anthropocene in critical geographic thought, and critical theory more broadly. Although both concepts are relatively novel, coming into prominence largely over the past 10 to 15 years, there are already certain tendencies that are beginning to calcify. First, there is a tendency to treat resilience as simply an outgrowth of neoliberalisation – a tendency that has begun to come under sympathetic critique from some quarters (Chandler 2014; Anderson 2015; Wakefield 2017; Grove 2018). Similarly, there is a tendency to embrace the Anthropocene as a provocation to move beyond modernity's confining spatial and temporal imaginaries – a tendency that has likewise begun to receive critical scrutiny (Colebrook 2012; Chandler 2018; Wakefield, 2019). This text seeks to highlight and amplify those 'minor' (after Deleuze and Guattari 1986) critiques, to ask what we might learn about our contemporary condition by juxtaposing those critiques with each other.

Just as this book attempts to move beyond dominant tropes on resilience and the Anthropocene, so too do we take a slightly different tack to our introduction. Rather than presenting a broad overview of core topics and summarising the contributors' essays, we have assembled here our collective reflections on three major themes inspired by our editorial engagements with the contributions. Thus, while this introduction is positioned at the opening of the collection, it is more a reflection on the major themes that, in our reading, emerge out of the essays. Accordingly, rather than creating an authoritative summary of each contribution, we offer here one tracing of how the essays speak to each other in provocative ways that go against the grain of conventional critical thought on resilience and the Anthropocene. Other tracings are surely possible, but our hope is that our reflections that follow on resilience, the Anthropocene and political subjectivity, respectively, will highlight key areas for further conceptual development as critical scholars continue to grapple with the challenges resilience in the Anthropocene poses to critical thought.

Resilience

To claim that the concept of resilience lacks clarity is, by now, a well-worn refrain among critical and applied scholars alike. Research from a variety of disciplines over the past two decades has detailed multiple contradictory and incompatible definitions of resilience that circulate within diverse policy and academic fields (Grove 2018). This has resulted in bipolar debates over the concept's political efficacy and pragmatic utility. On one hand, for many critical scholars, resilience is nothing more than the latest iteration of neoliberal governmental rationalities. Pointing to the way resilience initiatives often attempt to decentralise decision-making and fashion subjects capable of living and thriving with risk, these critics are quick to dismiss the concept on ideological grounds (MacKinnon and Derickson 2012; Watts 2015). On the other hand, for many applied (and some critical) scholars, resilience offers a potentially innovative approach to social and environmental governance, but its lack of conceptual clarity impedes practitioners' ability to operationalise the concept. Resilience thinking offers novel ways of integrating the social and environmental and increasing public participation and long-term planning in decision-making, but realising this potential requires more precise definitions of resilience to guide practitioners' reform efforts (Meerow et al. 2016). Thus, while ideological critiques know all too well exactly what resilience is, and can thus confidently dismiss the concept accordingly, application-oriented research does *not* know what resilience is, and thus seeks definitional clarity to realise the concept's value.

Our introduction, and this collection as a whole, does not attempt to adjudicate between these paradoxical readings. Instead, we seek to historicise the concept by positioning its emergence alongside the contemporaneous emergence of the Anthropocene. While resilience has circulated on the margins of fields such as engineering, psychology and ecology for decades (each, of course, with distinct and contradictory understandings of the concept), it began to gain prominence within policymaking circles during the late 1990s and early 2000s, as scholars and practitioners alike grappled with a series of social, geopolitical, technical and political economic events that exceeded modernist technologies of security premised on boundaries, prediction, stability, linear temporality and control. The end of the Cold War and the identification of non-traditional security threats, the United Nations Framework Convention on Climate Change's naming of dangerous climate change as a threat to development and well-being, the events of 11 September 2001 and their impact on national security planning, the conduct of warfare, and international financial and reinsurance markets, the 1998 Asian financial crisis, and increasingly catastrophic hurricanes, cyclones and typhoons throughout the tropics – to name but a few – formed the backdrop in the early

2000s against which Paul Crutzen, Will Steffan and other scientists began naming the Anthropocene as a distinct geological era. Since then, the Anthropocene has come to stand in for all manner of conditions that, we are told, reveal humanity's embeddedness within complex social, environmental and technical systems that threaten Earth's habitability.

Resilience became an increasingly influential governance principle alongside and through this growing recognition among scholars and policymakers alike that the stable, predictable environment many attributed to the Holocene, like the stable, predictable world of European modernity, were untenable assumptions. Its influence lies in the way the concept transvalues modernist security (Chandler, 2014). As Simon Dalby's chapter in this volume details, through his discussion of the links between earth systems science, the planetary boundary framework and thought on global security, resilience offers a theory of growth, development and improvement through *embracing* change, diversity, surprise and disruption, rather than banishing these conditions beyond the limits of the sovereign subject. It seeks the sources of security *within* the threatened object itself: it focuses attention inward, to life's systemic capacities for self-organised adaptation to external shocks, rather than outward to borders and bordering practices that attempt (and inevitably fail) to prevent disruption. Those spatial and temporal boundaries that attempted to purify this subject do not provide the socio-political conditions for development; instead, the modernist pursuit of stability, purification and control only increases the likelihood of catastrophic systemic collapse. Thus, at the moment the Anthropocene annihilates modernity's metaphysical fantasies of security-as-stasis, resilience arrives to reconstitute security as a problem of affirming rather than rejecting worldly connectivity and emergence.

For us then, the Anthropocene brings into focus resilience as a coherent body of thought. Our use of the term *thought* merits brief clarification. We follow Stephen Collier's (2009: 93) engagement with Foucault's College de France lectures to identify thought *not* as a 'passive response to discursive structures and power/knowledge regimes that define conditions of possibility for certain modes of understanding and acting', but rather as 'an as active response to historically situated problems ... [that] shape new technologies of power'. Collier's specification of thought gets around the thorny problem of trying to give resilience a coherent definition. Thought is not uniform, unitary or coherent across time and space. It cannot be deductively read off of a discursive regime, ideological project or ontological affirmation. Instead, it is a situated, creative and technically mediated practice of reflecting on the limits of the present – those material-discursive-technical elements whose arrangement constrains the possibilities for thought and action in some way – in relation to problematic situations. As Paul Rabinow (2011: 12) emphasises, thinking involves clarifying situations, and is oriented towards 'achieving a degree of resolution of what was problematic in the situation in the first place'.

In this light, resilience names a problem-space where critical reflection on the limits of the Anthropocene present is possible. Resilience thinking engages problematic situations that exceed modernist practices of security, such as problems of non-linear ecosystem change and collapse (Holling 1973; see Grove 2018). It offers a critique of modernist planning practices oriented around logics of centralisation, control and prediction as causing those environmental problems centralised planning sought to prevent in the first place, and instead offers a variety of governmental reforms designed to reconfigure social and environmental governance around principles of reflexivity, adaptive management and institutional change. Importantly, however, these practices of critique and intervention are not coherent across time, space or professional field. The critiques resilience thinking engenders can and do produce contradictory and incompatible styles of engaging with spatial interconnection and temporal emergence: while urban security practitioners present resilience as a problem of infrastructure hardening designed to prevent surprises – such as terrorist bombings – that threaten urban circulations, urban ecologists promote resilience as a means of living with and developing through emergent disruptions (Coaffee et al. 2009; Evans 2011).

Focusing analytical attention on these contextualised and situated practices of *thought* thus allows us to inductively examine resilience in the Anthropocene without relying on the deductive identification of the formal qualities of resilience. The formal similarities between resilience and neoliberalism – a common scepticism of centralisation, the production of risk-bearing subjects – matter less here than more subtle practices of critique that reconfigure multiple strategies, rationalities, techniques and practices of government in response to qualitatively novel experiences of time and space. The latter enable us to explore how resilience emerged out of specific historical situations, in response to specific problematisations of government engendered by social and environmental processes we now mark as 'the Anthropocene'. This is a key theme in the chapters in this volume from Sara Nelson and Kevin Grove and Allain Barnett. Nelson analyses C.S. Holling's early work on ecological resilience with the International Institute of Applied Systems Analysis to situate resilience within the wider trajectory of systems theory, and its systems–cybernetic governmentality. Grove and Barnett, in turn, demonstrate how the cybernetic behavioural science of Herbert Simon shaped influential resilience thinkers' understandings of complexity and adaptation. Both of these analyses demonstrate the 'environmental' qualities of resilience: that is, the way resilience initiatives operate through the problematisation and instrumentalisation of ecological relations. While resilience and neoliberalism may both mobilise environmental forms of power, these analyses each demonstrate, in their own way, how resilience is irreducible to neoliberal governmental rationalities. Nelson demonstrates how the modes of knowledge production associated with the systems sciences – and which Holling's work in ecology helped shaped – emerged through collaborations between scientists in capitalist and socialist countries, in which group scientists worked through contextually specific problems of the relation between markets, state planning and decentralisation. Grove and Barnett argue that resilience recalibrates the study of nature–society relations around a cybernetic will to design. Rather than attempting to reveal the objective (and thus predictive) truth of social and ecological phenomena, resilience initiatives focus on developing pragmatic and partial solutions to indeterminate problems of complexity.

As a body of thought, resilience thus opens on to a problem-space where social and environmental governance are reconfigured around cybernetic and designerly strategies and rationalities. These strategies become the pivot around which resilience plays with the limits of modernity, that series of inside/outside divisions that structure modernist understandings of self, subject, agency, the state, politics and expertise, to name but a few. For example, the problem for governance is no longer about developing predictive knowledge that enables human control of complex social and ecological phenomena, but rather how to work with and through the emergent 'environmental' powers of life itself (now given, in the Anthropocene, in terms of complexity). This is a problem of how to interiorise the exterior: how to recalibrate governance, politics and science for a world where emergence and interconnection make up the weft and weave of the bios. In the process, and as Lauren Rickards' chapter in this volume details, the boundaries that artificially separated the state, science and the public are becoming reworked through new practices of adaptive management and adaptive governance that design-in reflexivity to decision-making processes. Governance and expertise become more provisional here (Best 2014): not only do these practices horizontally redistribute authority and render technical expertise circumspect, they also expand governance to nominally include input and participation from end users formerly grasped as passive recipients of public service provision.

This play runs in the opposite direction as well. Even as resilience asserts that complexity decentralises expertise, redistributes authority and thus demands provisional and reflexive styles of governance, as Stephanie Wakefield's chapter in this book demonstrates, it *also* re-affirms a unified vision of the world – a 'one-world-world' of complexity. Imaginaries of world as a self-contained biosphere extend visions of coherence and harmony – modernity's visions of a stable, interior *bios* sheltered from a wider world of emergence – to the whole of planetary existence (see Fagan, this volume). This effectively exteriorises modernity's interior: faced with the Anthropocene's radical asymmetry between human and earthly powers, resilience affirms a coherent world of complexity that elides any form of division or asymmetry between the social and the natural, or within the social itself. And as Madeleine Fagan's chapter here explains, for ecologists, overcoming of the problems that complexity generates requires a radical escapism, the revolutionary transformation of this complex world into new, more sustainable configurations. The world itself becomes open to new forms of geo- and eco-constructivist interventions, new efforts to transform both macro-scale planetary systemic dynamics (geo-constructivism) and micro-scale eco-systems (eco-constructivism) in the name of resilience (Neyrat 2019).

Situating resilience in relation to the Anthropocene thus complicates the easy equation between resilience and neoliberalism that many first-cut critiques of resilience offered (Grove and Barnett, this volume). While there may indeed be formal similarities between the two, there are more nuanced differences and continuities that cast broader trends in contemporary critical theory in new light. For example, resilience scholars' affirmation of complexity as an ontological and biopolitical foundation for resilience initiatives, a move denounced by critical scholars for uncritically re-asserting a post-Cartesian ontology, mirrors recent critical affirmations of an ontologically prior power of life itself: both attempt to shore up their grounds for analysis and truth claims through recourse to an unassailable world of facts (Chandler 2014; 2018; Barnett 2017). And both respond to the loss of modernist grounds for truth: whether the affirmation of an objective world open to totalising, calculative rationality, or the affirmation of a world determined by objective class interests, both resilience thinking and new materialist thought grapple with the challenge of how to ground truth claims in the face of indeterminacy. Resilience proponents' affirmations of a world of complex systems and creative emergence can be read as mirroring that of decolonial scholars' affirmations of the pluriverse and its displacement of the 'one-world-world' (Law 2015; de la Cadena and Blaser 2018): both attempt to create space for multiplicity and difference; the world of complexity, like the world of the pluriverse, is a world that exceeds the knowledge of the individual and can only be grasped through partial, reflexive engagements with often irreconcilable difference (Wakefield, this volume). The distance between a critical ethic of care for difference and ontological multiplicity is thus not as far removed from neoliberal political economists' pragmatic ethic of institutional design as it might first seem (for examples of this designerly ethos, see Ostrom 1990; Ostrom 1997, Buchanan 1959; see Collier 2011; 2017 for overviews). Despite important ontological and methodological differences, both advocate an ethics of experimental and reflexive engagement with an indeterminate world. The recent critical embrace of experimental governance and experimental politics thus resonates in curious – and largely under-explored – ways with neoliberal affirmations of institutional design (see Rickards, this volume; Chandler 2018).

Situating resilience in the Anthropocene thus opens new conceptual, ethical and analytical challenges for both applied scholars and critical theorists alike. It points to subtle continuities between resilience thinking, certain new materialist and decolonial strands of critical thought, and neoliberal political theory that are irreducible to formal similarities, and instead play out on the more practical terrain of strategies and rationalities. The resonance between resilience and neoliberal theory is thus at one and the same time deeper *and* more contingent than critics of resilience allow. And yet, this recognition also highlights the limit of contemporary critical affirmations of life itself and ontological multiplicity. Like Hegel's Owl of Minerva spreading its wings with the falling of dusk, resilience becomes available to thought at the moment its historical condition passes away. But so too does the critical affirmation of a neo-vitalist ontology assert itself at the moment that the Anthropocene calls life itself into question (Neyrat 2019; Grove and Barnett, this volume). The question for us then becomes, what possibilities for thought does the Anthropocene hold out if we turn away from, rather than embrace, these ontological affirmations?

Anthropocene

As analysed above, the Anthropocene seems to bring the conceptualisation of resilience to the forefront. We believe that it does this through forcing advocates of resilience to think through how it works as a set of ideas and practices, in fact, as an ontology – a way of understanding the nature of being. What do we tap into when we think about the powers or capacities of resilience? There is little doubt that resilience has an ontology of life as a resource to be drawn upon. How is it that life is a resource? We do not mean that life is a resource as in a pile of materials that can be extracted or worked upon through new more productive technologies but rather that, for resilience theories, life is necessarily always in excess of being: i.e. that there is always untapped potential in the here and now. Resilience as a set of policy practices is therefore oriented towards enabling this potential to come to the surface, to circulate or emerge. In this way, feedback effects necessary for complex self-adaptive systems to operate efficiently are seen to enable adaptive transformative effects, 'bouncing forward' rather than merely 'bouncing back' to a previous equilibrium.

This first point is crucial to grasp. In high modernist approaches to 'development', 'security' and 'progress' there is a strict subject/object or human/nature divide. Humanity is the creative agent or actor and the world/nature/nonhumans are merely passive objects of timeless universal causal laws. Human agency is the driving force for creativity and change and the nonhuman world awaits to have its secrets unlocked through the application of modern science and technology. Resilience approaches seek to disrupt this divide, reallocating more creativity and agency to the side of nature/the nonhuman world. Life is in excess of being because life is itself the possessor of creative powers. In a world where it appears that the application of human science and technology to control or direct nature has undermined natural processes of regulation – including the catastrophic unintended consequences of climate change and global warming – resilience as a dominant policy framework seeks to slow down this run-away process by restoring more power to nature or life itself and seeking alternative ways forward that redistribute understandings of agency.

It is because resilience thinking searches for a solution in the hidden or potential processes, inter-relations and interactive emergence of life itself, that its 'designerly' approach is less concerned with 'top-down' interventions – seeking to impose directions and ends – and more with 'facilitating', 'enabling' or 'engendering' existing powers and capacities or seeking to redirect them to new possibilities. Perhaps resilience could be seen as a search for new powers within life itself, but life understood as potential, as virtual or immanent. It is here that life can be considered in excess of being, or as the virtual in excess of the actual, enabling new adaptive possibilities to emerge. We suggest that this way of grasping resilience as a form of biopolitics – a form of a politics directed to life itself – is crucial, but to bear in mind that the form of life which is being 'scaled-up' or 'engendered' is virtual or potential rather than actual or existential. Resilience approaches tend to shy away from a biopolitics of the actual: the use of high tech or scientific approaches of 'eco-modernism' to engineer or manage life in a more traditional 'top-down' or 'human-centred' way.

Resilience interventions seek to work indirectly, on seeing signs or feedback effects and enabling more flexible responses, rather than intervening directly to attempt to tackle causes or engineer outcomes. For example, where new technologies are applied or rolled-out more ubiquitously they are generally designed to see relational interactions and effects so adaption can be increasingly 'real-time' rather than to work on the world directly. It is because resilience approaches focus on the potential or the virtual powers of life that they always pay attention to what lies beyond the surface of appearances – beyond the world of entities and actual existence – to the hidden relations that are behind these manifestations. For resilience approaches then, nothing is exactly how it appears and entities are not reduced to fixed essences. Instead, relations are key rather than actual entities – entities as given to the subject, that can thereby be measured or counted.

It is because resilience approaches focus on the potential powers of life, already real but not actualised, that the key to policy interventions is system or relational understandings. Resilience as a framework of knowledge is therefore concerned with the way relational context shapes outcomes rather than with forms of universal or linear causality that can be generally applied. Thus relations are key to resilience as it is relational interactivity, which enables potentials to become actualised – to emerge. For resilience thinking, in some contexts particular entities may play a positive role in another a negative role: there can be no generalisations or 'one-size-fits-all'. A good example of a resilience ontology is provided by Nobel prize-winning economist Amartya Sen, who argues that absolute levels of education, wealth or poverty say very little about the quality of life of an individual, as these forms of representation always fail to grasp the relational context that enables capacities to become actualised (Sen 1999; Chandler 2013; Reghezza-Zitt and Rufat 2019). The resilience of an entity – an individual or a community – is always understood as a relation: the resilience of A to B. Resilience can thereby never be reduced to a property of an entity. It is thus the privileging of relations over entities that is so important in cybernetic understandings and the work, highlighted in a number of chapters, by C.S. Holling on ecological resilience and system maintenance and change.

The potential or 'virtual' power of life is necessarily relational because it is not embodied in entities but in their systemic interaction. Living systems are understood to have creative or 'emergent' powers – 'system effects' – that cannot be accessed by examining the properties of entities themselves. In other words, 'the whole is always more than its parts', in contrast to reductionist and atomised understandings. It is the powers unleashed by system relations that need to be understood, accessed and redirected and repurposed. It is for this reason that cybernetic approaches, which focus on cognitive or communicative interaction between parts in a system, whether human or machine, are so important to the trans-disciplinary reach of resilience thinking as an approach to understanding life's generative and productive capacities. This turn to the 'power of life' poses a clear challenge to modernist separations of human/nature and subject/object, paying attention to human-nonhuman systems and assemblages and their emergent and unintended effects. Thus, following Colebrook (2010: 32) we would argue that resilience thinking tends to emphasise three aspects that takes it beyond modernist or liberal approaches to governance:

*De-centredness – for resilience approaches there is no key or dominant defining attribute or underlying cause. There is no metric for resilience, which is why in the policy literature there is always a concern for how resilience might be measured. The size of a state's gross domestic product (GDP) or the amount of profits generated by a firm in the last quarter are no indication of these entities' resilience to shocks or unexpected events, in fact, often quite the opposite.

* Immanence – there is no transcendental 'secret' or ultimate mechanism or cause that would enable human direction and control over the power of life, life is seen as self-governing or self-producing, through relational interaction. Here, resilience thinking draws very much upon cybernetic and complexity understandings that the self-organising power of life, through small repeated iterations over time, has produced the richness and diversity of the world we live in. Life appears to be self-directed but not to be arbitrary or pointless and, in the Anthropocene, human attempts to impose our own will or direction upon these processes appears to be counterproductive and hubristic.

* Affectivity – resilience approaches stretch beyond merely human societal interactions, consciously bringing in human or social and technological relations with other nonhuman entities and forces, to consider system effects and the unintended consequences of human actions. For this reason, there is less emphasis on what are considered to be specifically human attributes of thought and cognition. The concern is much more with the powers and capacities that are brought into play through relation, thus attention is placed upon the particular affordances and sensitivities of entities, which are expressed in concrete contexts: the power to affect and to be affected.

For us, then, there are two key assumptions that enable resilience approaches: 1) life has to be in excess of being, and 2) relations must be privileged over entities. These key assumptions mark a distinct challenge to modernist or liberal universal frameworks of thought and clearly emerge in response to new problems and new sensitivities of the limits of modernity in the second half of the twentieth century. After having emerged to prominence in the last two decades, resilience would seem well placed to be a dominant discursive governmental framing in the epoch of the Anthropocene. However – as the chapters, which follow, analyse – a lot hangs on how we understand these two assumptions.

Life in excess of being

Firstly, the assumption that life has to be in excess of being is, on one level, straightforward; humanity is self-evidently not the only creative or agential being or we could not exist in the first place. Life is clearly in excess of being as evolutionary change demonstrates. Even non-life is in excess of being as the production of life from non-life illustrates (Povinelli 2016). There is little doubt that life harbours multiple potentialities only some of which are actualised. The assumption of resilience thinking, that we are now 'after Nature' (Purdy 2015; Lorimer 2015) or 'after ecology' (Morton 2009, 2013; Latour 2004), in that nature should no longer seen as a distinct and separate realm, as a mere 'background' for human struggles, is nearly universally accepted. Yet it is not so obvious that human and natural forces should be seen as mutually entangled and mutually co-constitutive. Perhaps the desire to remove the distinction can be seen as a consequence of critical social and political thought's desire to ensure that nature is not seen as some sort of stable ground for modernist hierarchies and exclusions (Clark 2011: 25). It may seem straightforward that after nature, the social and the natural should be co-constitutively grasped together. However, for many Anthropocene thinkers, the fact that life withholds excessive powers of actualisation does not immediately equate to these powers as being generative and productive 'for us', nor to the possibility that humans can have the capacity to understand, instrumentalise or to actualise these powers.

The problem is that, in the constructivist desire to 'de-naturalise' policy understandings, there is little separation of the human from the world, said to require constant adaptive management (Neyrat 2019: 82). In resilience discourses it can easily appear that human ingenuity is freed from the limitations of nature and that nature is now available for us through the process of its accidental 'humanisation' in the course of industrialised modernity. This new 'culture-nature' collective appears then to be the product of our construction and thus open to alternative constructions through the systems-thinking approaches of resilience. Thus, while modernity can be seen to have failed in its hubristic designs upon the earth, resilience can enable a new set of 'posthuman' governing techniques and frameworks open to the emergent powers of systemic relations.

As Nigel Clark, among others, has long argued, these framings of 'hybrid nature-cultures' all too easily pass over the excessive and unpredictable powers of nonhuman or inhuman nature. Rather than overcoming the nature/culture divide, some approaches within Anthropocene thinking therefore cast resilience perspectives as very much part of the modernist desire to see everything through a human-centred or anthropocentric lens:

> What we need to keep an eye on here is the repeated insistence that there is no outside to the new hybridised environments: thus no functionally intact nature enduring beyond, beneath, amidst or after this assimilation … It is a fusion, I want to argue, which discourages any political or ontological investment in a geo-physical materiality with an autonomy and integrity of its own.
>
> *(Clark 2011: 11)*

While Clark shares a similar perspective of interactive becoming or emergence, as that of resilience approaches, the difference is that the potential for taming these powers and putting them to productive use is questioned. The reason for this is that agential life – even if it is understood as a complex adaptive system of emergence – is not a mutual collaborative product to be put to use for human convenience. The vitality of matter is well beyond human knowledge and control, reaching down to the molten core of the planet and up to the impacts of solar

winds and radiation. The relationship between humanity and the planet is fundamentally asymmetrical, well beyond resilience thinking's imaginaries of 'mutuality
or co-dependence' (Clark 2011: 46). Authors, like Eugene Thacker, have also
emphasised this 'darker' or 'negative' side to immanent thought, which is lost in
resilience approaches, which concern specifically life *for us* as human beings
(Thacker 2010: xv). Anthropocene considerations of life – as 'inhuman' (Clark
2011), 'nonhuman' or 'unhuman' (Thacker 2010) – therefore point beyond the
productive biopolitics of resilience. Key to this shift is the distinction between
immanent understandings of life as contingent and interactive (shared with resilience thinking) and the metaphysical assumption, necessarily underlying resilience
thinking, that life has a positive, thermodynamic or Neoplatonic, telos or flow
towards creative differentiation (an understanding of emanence, radiating out from
either a theological or cosmic source) (Thacker 2010: 217).

Claire Colebrook argues that the inaccessibility of life's excessive potential leads
to thought beyond the ontological constraints of positive and productivist understandings of a 'redemptive' or 'knee-jerk' vitalism so often underlying resilience
thinking (Colebrook 2010: 48). It is too simplistic to imagine life as a force that
flows through relational interaction, as 'an end that unfolds through time' (2010:
22), that seeks to draw out essences or enable entities to 'become themselves' or to
orient themselves more productively to the world. The power of excess in the
pragmatic framings of resilience is always conveniently cast productively and functionally, where the power of life enables entities and systems to develop their own
internal principles for mutually adaptive forms of self-maintenance or autopoiesis
(2010: 34) – bouncing back to equilibrium or forwards to new forms of mutual
sustainability. Life is thereby reduced to the on-going work of survival and adaptation. Thus, although the power of life may have no human-centred liberal telos
of progress, in constructions of resilience, life is always amenable to functional
collaborations of mutual survival and sustainability. This is also reflected in much of
contemporary social and political thought, for example, in Bruno Latour's imaginaries of collective assembly and negotiation to construct with nonhuman others
who share our 'Earthbound' existence and the 'compositioning' and 'companion
species' of Donna Haraway (Latour 2018; Haraway 2016; Clark 2011: 36–40;
Neyrat 2019: 90–104).

In tracing a line of thought of a less productivist and activist framework of the
power of life, life is not imagined as continually working to become its 'better self'
through the imaginary of the 'hidden hand' of resilience. For Colebrook, these
imaginaries set 'the urgent, yet redemptive, tone today of ecological ethics', and
constitute resilience thinking as part of the problem rather than the solution as: 'it is
the insistence on the universe as an organism or web of life that allows us to retain
anthropomorphism, for the world is still the milieu of our life *and* life itself is presented as active, creative and self-furthering' (Colebrook 2010: 57, italics in original). Along similar lines to the Anthropocene thinkers above, she argues that we
need to reject this view of life as made up of systems of harmonious self-making
interactive subjects and instead to appreciate that to live is also to become subject

to powers beyond knowledge and control (2010: 133). It is precisely these breaks in continuity that prevent life being one homogenising process of 'becoming' or 'actualisation' and enable creativity beyond the biopolitical imaginaries of resilience. While resilience thinking challenges modernist assumptions of human-centred direction over life, anthropocentrism is smuggled back in with an ontology of a world that is coherent and harmonious and capable of directing governance towards new forms of sustainability.

Anthropocene thinkers tend to stress that life lacks an immanent direction towards order or functional individuation and differentiation. What appears as a telos, particularly to reductionist forms of modernist thought, is merely a product of contingent interactivity: life may have infinite forms of hidden potential but that does not mean that it is equipped with an underlying 'purpose' or 'reason' or a vitalist force that can be tapped into and directed. To think so would be merely to reproduce (in a slightly more mediated way) discredited assumptions of linear causality, recapturing imaginaries of 'progress' and 'development' for new (seemingly less modernist) forms of governmentality. For Timothy Morton, for example, the Anthropocene reveals a 'Dark Ecology' at work, where the potential excess of life over being is all too real but ultimately, ontologically, inaccessible to us. Rather than readable and adjustable feedback loops, so essential to discourses of resilience and adaptation, there is a fundamental, irreducible, gap between effects or appearances and things or entities themselves (Morton 2016: 93). For Anthropocene thinkers, the fact that life is in excess of being poses fundamental questions to resilience discourses that seek to use or instrumentalise life as a resource. Resilience approaches seem just as hubristic and blind to unintended consequences as those of the modernist framings they seek to go beyond.

Relations over entities

The second assumption, that relations should be prioritised over entities, is one that appears to be even more problematic in the Anthropocene. This framing should not be confused with what Deleuze understood to be a fundamental breakthrough in philosophy, that 'relations are always external to their terms' (Kleinherenbrink 2019: 51). In fact, the implications of privileging relations are quite the opposite. Deleuze, along with some later speculative realist and object-oriented approaches, sought to highlight that all entities contained a virtual side, inaccessible through appearance or effects (Colebrook 2010; Kleinherenbrink 2019). In resilience approaches, time and again, entities – in the case of resilience policymaking, individuals, communities and societies held to be 'vulnerable', 'at risk' or 'failing' – are seen as products of their relations. This fundamentally reverses the view that entities are inaccessible, reducing them instead to mere ciphers or signs of their relations. Problems are then understood as products of 'maladaptation' or poor choice-making, as a result of particular relational 'path-dependencies' which are then capable of being traced, mapped and reordered, often through external 'capacity-building' governmental expertise. As with the first assumption, of life in excess of being, the prioritisation of relations over entities understands outcomes as always 'co-produced', equalising or redistributing agency away from the more hierarchical framing of modernist being which puts more emphasis on fixed attributes or capacities.

Resilience thinking is about the availability of the relational interconnections of which entities or surface appearances are an indication or sign. Therefore the importance is the context in which relations are to be understood as productive and therefore adaptable and the need for different relational understandings. Resilience is not about adding something that does not exist to this context – for example, in a modernist or liberal perspective, poverty or vulnerability might be addressed by providing aid and resources – but about working indirectly via relations rather than directly, upon entities themselves. A relational understanding of resilience would argue that there is no essential link, for example, between poverty and resilience; it is possible that poor people can adapt better than rich people to particular crises or problems. It is the particular relation that is key to adaptive capacities rather than the entity itself. Entities in-themselves, i.e. at the level of appearances, of the already actualised, can tell us little about the virtual or potential capacities that resilience seeks to draw from them via their relational entanglements.

It is for this reason that policy areas dominated by resilience thinking, such as disaster risk reduction, focus much less on the provision of resources and instead on capacities for resilient forms of adaptation through reimagining relations both inside the entity/community and with the external environment. As the United Nations programme for sustainable urban development argues:

> Though there are shared characteristics, 'poverty' and 'vulnerability' are not the same thing. While poverty reflects a lack of economic and social assets, vulnerability additionally implies a lack of capacity, security, and exposure to risks. Though the overlap is significant, not all poor are vulnerable and not all who are vulnerable are necessarily poor. This has important implications for policy – as does understanding the assets and capabilities even very poor populations possess in their resilience and response to either slow-onset climate change or disasters. Much can often be built from communities, especially once assumptions regarding their capacities are put aside.
>
> *(UN-Habitat, 2014: 15)*

Resilience thinking then works on the unseen, the virtual, or the potential of a community or entity's relational interactions rather than the actual or measurable attributes of the entity per se. This can perhaps be seen as a 'flatter', a more equal or 'democratic' approach which is no longer in thrall to traditional hierarchies and state-based forms of representation – much as 'human security' or 'human development' indexes seek to challenge modernist accounting on the basis of universal GDP statistics. This can perhaps be seen as major weakness, inherent to resilience framings. The problem is that these 'flatter' forms of understanding can be profoundly depoliticising, not only when it comes to assuming a mutual co-

constitutive relation between humanity and natural forces but also when entities are seen as equally productive of their circumstances and when contexts are often reduced to relations between societal or community forms and external threats or risks. In resilience framings, relationality is articulated in terms of a community or society context and the enabling of pragmatic responses to particular threats or crises seen as unpreventable or inevitable. The effects are thus understood to be a product of relational context, i.e. to be non-linear in outcome, depending on specific relational adaptive capacities.

Anthropocene thinking provides a critique of the resilience assumption that shocks are 'inevitable' and that adaptation is the only way forward, understanding relationality at the level of causes as well as of effects. The Anthropocene takes its name from the recognition that climate change and other environmental problems are a product of human activity. The crisis or threat is not external to us: the threat to communities' ways of life is not some purely external force but one that has been in many ways anthropogenically created. This immediately problematises the approach of adapting to save or preserve modern forms of production and con-sumption. By contrast, in discourses of resilience, the problems, shocks and instabilities being responded to are always constructed as 'inevitable', in a complex or 'non-linear' world where life is much less predictable. This framing constructs resilience as always a process of adapting to external or outside forces and resilience policy interventions then concern the hidden societal or community relational capacities for internal or self-organisation, deploying different technologies and approaches to change, learn and adapt so as to enable social progress towards the liberal normative goals of 'peace, rights and development' (Tocci 2019).

As Frédéric Neyrat (2019: 78) notes: 'The ecology of resilience has so completely accepted the axiom of turbulence that it finds itself in the situation of being onto-logically incapable of giving an account of the turbulence that nourishes it.' Thus the focus on relational effects and the need for adaptation necessarily suborns politics to the governance of effects rather than causes. In this framing, the Anthropocene of global warming and climate change is always necessarily something external, some-thing constructed as happening *to us* and in the face of which we must become resilient, accepting our fate of coping on the edge of crisis. Ironically, then, resilience approaches seem to be the enemy of resilience in the Anthropocene, striving to continue to preserve and reproduce exactly the same modes of being and thinking that caused anthropogenic climate change in the first place. Anthropocene thinking does not just flag up the problems of resilience assumptions regarding the pro-ductivity or instrumentalisation of life in excess of being, it also profoundly questions this depoliticisation of crises and coping mechanisms and the redistribution of adap-tive responsibility on to communities that are already marginalised. Liberal goals and understandings are no longer seen as positive but as part of the problem that needs to be overcome. This changes everything, as Naomi Klein (2014) might say.

Thus the Anthropocene appears to reveal two major sets of gaps or aporias, which are fundamental to resilience thinking. Firstly, resilience discourses necessa-rily assume that, because life is in excess of being, it has a productive potential that has been untapped in modernist frameworks, which see life as passive rather than possessive of creative powers. This assumption that life's withdrawn or excessive potential is there 'for us' has been fundamentally challenged in the Anthropocene. If anything it is the unrecognised excessive and immanent powers of life beyond human finitude that are revealed to us in our contemporary epoch precisely as uncontrollable and unknowable. It is these potentially infinite powers, which appear to dwarf human imaginaries and call into question further attempts to extract and instrumentalise nature as 'there for us'. Secondly, in resilience discourses, the complexity of unseen or unrecognised policy-feedbacks is rarely framed to take into account the structural causes of resource depletion and social and economic marginalisation, which makes communities and societies more vulnerable to shocks and disturbances (Neyrat 2019; Whyte 2019). As a consequence of this framing, the policy interventions, based upon enabling and capacity-building for resilience, often tend to impose the costs of adaptation upon those who are already in a marginal position, attempting to tackle the effects of crises and shocks but never the structural causes (Moore 2019; Chandler and Reid 2019).

Another way of looking at these two aspects is that Anthropocene thinking extends and transforms the focus on excess and relationality of resilience thinking in ways that could be read as pushing further away from modernist or liberal binary assumptions of a nature/culture divide but at the same time realising that agential becoming cannot be reduced to a totalising or homogenising process. Rather than seeking to combine or harmonise human actions and the processes of the physical environment – in effect, reducing the politics of life to the immediate appearances of the world and pragmatic responses to these – Anthropocene thinking emphasises the clash of temporalities and scales of processes and interactions in ways which attempt to give their due to both human and nonhuman forms of agency.

Political subjectivity: the end of the world and after

In tracing these lines of contact between resilience and the Anthropocene, this book is perhaps the first to seek to show how the end of the world works politically: as the end of differentiations. First there is the end of modernist political divides between right/left that resilience represents, then the end of modernist metaphysical divides (human/world, subject/object) that the Anthropocene critique of resilience represents. With this in mind it is worth reflecting on the meaning and experience of indifferentiation in a concrete context. After all it is easy for theorists to state that the Anthropocene overcomes or disrupts the human/nature divide but they have so far failed to show how this works in terms of real life subjectivities. Scholars typically use empirical examples which literally show the interactions of humans and nature and impact of climate change, or they do philosophy and seek to make metaphysical claims about the nature of being, most often drawing on Continental philosophy. Neither approach however helps us to understand changes in political subjectivity that the turn to indifferentiation and immanence entails.

How to understand the political subjectivity of immanence/indifferentiation? Does the end of relation and turn to indiffentiation enable new possibilities and political openings? Is it a matter of staring into the abyss without fear? Or perhaps more darkly, is life as immanence an end without opening, promising only more indifferentiation?

Philosophers and political theorists have long imagined that immanence might 'save' us, imagining it as a proper ontology counterposed to modernity's long history of separation (dividing the world into subjects and objects, for example, or an outside human species to order the world). In a similar way, insofar as Anthropocene critical theory and resilience thinking alike both promote human-nonhuman entanglement, and ways of dethroning what is now seen as the outdated hubristic humanist subject via enmeshment into complex systems, Anthropocene thought, we might say, also proposes immanence as salvation.

But as Jean Luc Nancy (Nancy and Barrau 2015) writes, late liberal societies have *already* arrived at immanence, however not as a liberating pathway out of the Anthropocene. Instead immanence is a description of the kinds of life now lived amidst the Anthropocene's ever-spiralling forms of eco-cybernetic control and connection. This includes not only transformations occurring in the realm of governance, capital, logistics and militarisation, but also the existential transformations unfolding now via ubiquitous social media, weaponised online networks and information saturation. Here the deterritorialisation that this indifferentiation implies is not just a matter of what capital does concretely to people and things (primitive accumulation, displacement, uprooting of tradition and lifeways, reterritorialised into working classes, factories, cities ...). It is also a matter of the psychic and subjective dimensions of this. As Gilles Deleuze and Felix Guattari write in *Anti-Oedipus*, 'capitalism tends toward a threshold of decoding that will destroy the socius in order to make it a body without organs and unleash the flows of desire on this body as a deterritorialized field' (Deleuze and Guattari 2003: 36).

Far from salvation, the living immanence/indifferentiation taking form (or nonform) across the realms of infrastructure, computing, algorithms, social media networks, disasters and advertising agencies is instead experienced as a kind of hypersaturation or nondifferentiation. Attempting to imagine what it would be like to actually experience pure immanence, philosopher Quentin Meillassoux suggests imagining:

> What our life would be if all the movements of the earth, all the noises of the earth, all the smells, the tastes, all the light – of the earth and of elsewhere, came to us in a moment, in an instant – like an atrocious screaming tumult of all things, traversing us continually and instantaneously. As if the nothing of death could not be understood as a simple void, but on the contrary only as a saturation, an abominable superfluity of existence. Death, thus understood, is the triumphant reign of communication. To die is to become a pure point of passage, a pure centre of communication of all things with all things.
>
> *(Meillassoux 2007: 104)*

Amidst the now-time tangle of feedback and communication, notifications and pop-up alerts, when everything and everyone – even, increasingly, in many 'smart' projects, nonhumans – is transmitting/receiving, there is less selection of images, and more an unceasing algorithmically choreographed barrage, a colliding, amassing heaping, where everything becomes equivalent (Nancy and Barrau 2015; Nancy 2015). Resilience hopes to manage these connections; Anthropocene theorists celebrate them; much of the population endures them. Within such entanglements human subjectivity becomes less any kind of liberal subject, and more a point of pure passage or transmitting node. For Meillassoux this immanence is a kind of death, one that occurs 'by dissipation of the body, by an ever-wider opening of the latter onto the external flux, up to a complete dissolution' (Meillassoux 2007: 103). This, we might imagine, is also the space of the Anthropocene, at least one dimension of experience in it. If this is the case, it challenges contemporary celebrations of immanence and relationality both those of resilience and of Anthropocene critical theory.

Rather than further celebration of indifferentiation, what is needed, another line of Anthropocene thinking might suggest, is an anti-productivist, anti-relational, and anti-equivalence worldview – and politics. But for this to be possible requires separation and detachment: from the entangled relations which are said to define humans and nonhumans, and from the suffocating immanence of late capitalist networks, resilience governance and Anthropocene theory alike. As Neyrat (2019) argues, there could be no conditions of possibility for politics if it were ontologically impossible to stand apart or separate from the flux or flow of the immediacy of life processes. This circles us back to the passage cited earlier by Nigel Clark:

> What we need to keep an eye on here is the repeated insistence that there is no outside to the new hybridised environments: thus no functionally intact nature enduring beyond, beneath, amidst or after this assimilation ... It is a fusion, I want to argue, which discourages any political or ontological investment in a geo-physical materiality with an autonomy and integrity of its own.
>
> *(Clark 2011: 11)*

From this perspective, rather than immanence and indifferentiation we have asymmetry and autonomy as a basic starting point. The world is not a flat, endlessly entangled 'collective' (Latour 2018) amenable at all times to control or knowledge. Instead, there are indeed insides and outsides; realms completely unexplored by humans – Earth's molten core, for example (Clark 2018) – and others which resist being made productive altogether. This, we might add, is true of the geo-physical world and for human beings as well – a point well detailed through contributions to this volume from Sébastien Nobert, Harold Bellanger Rodríguez and Xochilt Hernandez, and Nigel Clark. Nobert and colleagues detail through a study of living with slowly-seeping volcanic air pollution in rural Nicaragua how disaster resilience initiatives attempt, and largely fail, to limn the explicated temporality of resilience planning with the deep geological time of the volcano. For Clark, paying attention to the life-conditioning dynamics of the inner planet destabilises dominant understandings of vulnerability as a social effect, for it exponentially increases the asymmetrical relation between planet forces and earthly life. Thus, acknowledging that the world is not there for us – that it has its own autonomy, separate from one's own – requires giving up forcing definitions on it. Maybe 'life' is vital, emergent, productive. Maybe it is not. Maybe 'the world' can be put to work in manifold ways. Maybe not.

But just as equally, acknowledging that the world is not there 'for us', that it has its own autonomy, need not lead to meditations on human powerlessness. What we have outlined in the preceding sections of this introduction is a new critique of resilience. The latter's problems are not so much that it is neoliberal: it could just as easily be radical or communist or postmarxist, and indeed the ontology undergirding each of these is often very similar to that of resilience. Instead the Anthropocene reduces the political differences that were important for modernity – if the world is there for us, then we will continue to destroy it and if it is not there for us, then we have to destroy ourselves. Claire Colebrook's chapter in this volume starkly details this point. In her reading, resilience emerges through a dialectic that positions reason and whiteness as both proper and natural, on the one hand, and fragile and constantly threatened, on the other. But the spread of (white) reason – even in current cybernetically inflected 'bounded' variants – occurs through the negation and denial of alterity, the eradication of life it strives to secure. From this perspective, we cannot have the human and the world in the Anthropocene. Resilience merely works on shifting the balance between human and world and works on this as a continuum. Some aspects are more human (intervention – capacities to know and datafy and respond to effects cybernetically) some aspects are more world (non-intervention – enabling natural forces to be creative, like markets/evolution/complex adaptive systems). Resilience seeks to rework the human/world divide while maintaining the categories, while the Anthropocene begins to raise problems with whether this can make sense or address the problem.

Anthropocene theory tends to emphasise the latter as antidote to what many authors see as the problem of modern hubristic humanism. As David Chandler outlines in his chapter for this volume, affirmations of the Anthropocene assert that modernist tropes of technologically mediated, progressive separation of the human from the natural ignore that this progress was steadily destroying the possibility of planetary survival. In the Anthropocene, there is no 'outside' from which resources could be drawn, or which offers the promise of revolutionary redemption over its horizon. The liveliness of weeds versus the humbleness of the human is perhaps the quintessential Anthropocene theory trope (Chandler and Reid 2019: 74–5). But even if the world is not there for us, if it has its own autonomy, do we, as humans, in fact have to destroy ourselves? Or are there other possibilities beyond this false binary to which resilience and Anthropocene thinking often attempt to tether human being?

The power of earth processes can, from another vantage point, be understood as an incitement, a provocation, a gift. In our view this is precisely the crucial point Nigel Clark drives home in his chapter in this volume. While emphasising the asymmetry of earth and human powers, the conclusions Clark draws from this fact seem to us to diverge radically from most Anthropocene thinking on the same matter. That much of the earth is unknown or uncontrollable does not require as its corollary that humans sink into self-hatred or disavowal of our own capacities. Instead, as portrayed by Clark, this fact represents part of the beauty and tragedy of living, a context within and sometimes against which deeply varied human endeavours are forged. Wakefield (this volume) emphasises that it is urgent that we open up these possibilities against the closing down of experimentation, either under the rubric of resilience, attempting to ward off the Anthropocene, or under discourses of affirmation, with their suffocating uniformity, seeking to disabuse us of human powers and capacities.

As we see it, there are living and nonliving processes beyond our knowledge and control but there are also real histories, not just of the impact of the human 'species' on the environment but also of conscious political subordination, contestation and struggle, which also have all too real effects which continue to reverberate no less than those of pre-human planetary forces. At a time when human power and hubris have become objects of ubiquitous scorn for Anthropocene thinking, it is more important than ever to remember the varied forms that human impact actually takes. The exploitation of workers from factories to call centres in the name of profit, the boxing of whole populations into prisons, and so on. But just as equally human impact includes great proletarian struggles and tactics developed therein from sabotage and strike to insurrections and mass uprisings for freedom. The American Civil Rights movement or the 19th and 20th century workers' movements, to name just two examples: each is singular – nonequivalent – in its own right, each waged by political subjects that separate themselves from contexts and relations which strangle them. Such movements are not just objects of remembrance but human forms elaborated in ever-new ways to this day, as recent developments in Hong Kong testify. We have little doubt that this provides a much richer ontological perspective than reducing appearances to unknowable and arbitrary processes and politics to reactive adaptation.

References

Anderson, B. (2015) What kind of thing is resilience? *Politics* 35(1): 60–66.

Barnett, C. (2017) *The Priority of Injustice: Locating Democracy in Critical Theory*. Athens: University of Georgia Press.

Best, J. (2014) *Governing Failure: Provisional Expertise and the Transformation of Global Development Finance*. Cambridge: Cambridge University Press.

Buchanan, J. (1959) Positive economics, welfare economics, and political economy. *The Journal of Law and Economics* 2: 124–138.

Chandler, D. (2013) 'Human-centred' development? Rethinking 'freedom' and 'agency' in discourses of international development. *Millennium: Journal of International Studies* 42(1): 3–23.

Chandler, D. (2014) *Resilience: The Governance of Complexity*. London: Routledge.

Chandler, D. (2018) *Ontopolitics in the Anthropocene: An Introduction to Mapping, Sensing, Hacking*. Abingdon: Routledge.

Chandler, D. and Reid, J. (2019) *Becoming Indigenous: Governing Imaginaries in the Anthropocene*. London: Rowman & Littlefield.

Clark, N. (2011) *Inhuman Nature: Sociable Life on a Dynamic Planet*. London: Sage.

Clark, N. (2018) Bare life on molten rock. *SubStance* 47(2): 8–22.

Coaffee, J., Murakami Wood, D. and Rogers, P. (2009) *The Everyday Resilience of the City: How Cities Respond to Terrorism and Disaster*. New York: Palgrave Macmillan.

Colebrook, C. (2010) *Deleuze and the Meaning of Life*. London: Continuum.

Colebrook, C. (2012) Not symbiosis, not now: why anthropogenic change is not really human. *Oxford Literary Review* 34(2): 185–209.

Collier, S. (2009) Topologies of power: Foucault's analysis of political government beyond 'governmentality'. *Theory, Culture & Society* 26(6): 78–108.

Collier, S. (2011) *Post-Soviet Social: Neoliberalism, Modernity, Biopolitics*. Princeton: Princeton University Press.

Collier, S. (2017) Neoliberalism and rule by experts. In V. Higgins and W. Larner (eds), *Assembling Neoliberalism: Expertise, Practices, Subjects*. New York: Palgrave.

de la Cadena, M. and Blaser, M. (eds) (2018) *A World of Many Worlds*. London: Duke University Press.

Deleuze, G. and Guattari, F. (1986) *Kafka: Toward a Minor Literature*. Minneapolis: University of Minnesota Press.

Deleuze, G. and Guattari, F. (2003) *Anti-Oedipus: Capitalism and Schizophrenia*. London: Continuum.

Evans, J. (2011) Resilience, ecology and adaptation in the experimental city. *Transactions of the Institute of British Geographers* 36(2): 223–237.

Grove, K. (2018) *Resilience*. Abingdon: Routledge.

Haraway, D. (2016) *Staying with the Trouble: Making Kin in the Chthulucene*. Durham: Duke University Press.

Holling, C.S. (1973) Resilience and stability of ecological systems. *Annual Review of Ecological Systems* 4: 1–23.

Klein, N. (2014) *This Changes Everything: Capitalism vs. the Climate*. London: Simon & Schuster.

Kleinherenbrink, A. (2019) *Against Continuity: Gilles Deleuze's Speculative Realism*. Edinburgh: Edinburgh University Press.

Latour, B. (2004) *Politics of Nature: How to Bring the Sciences into Democracy*. Cambridge, MA: Harvard University Press.

Latour, B. (2018) *Down to Earth: Politics in a New Climatic Regime*. Cambridge: Polity Press.

Law, J. (2015) What's wrong with a one-world world? *Distinktion: Scandinavian Journal of Social Theory* 16(1): 126–139.

Lorimer, J. (2015) *Wildlife in the Anthropocene: Conservation after Nature*. Minneapolis: University of Minnesota Press.

MacKinnon, D. and Derickson, K. (2012) From resilience to resourcefulness: a critique of resilience policy and activism. *Progress in Human Geography* 37(2): 253–270.

Meerow, S., Newell, J. and Stults, M. (2016) Defining urban resilience: a review. *Landscape and Urban Planning* 147: 38–49.

Meillassoux, Q. (2007) Subtraction and contraction: Deleuze, immanence, and matter and memory. *Collapse* 3: 63–107.

Moore, J. (2019) Capitalocene and planetary justice. *Maize* 6: 49–54. Accessed at: https://www.academia.edu/39776872/The_Capitalocene_and_Planetary_Justice?auto=download.

Morton, T. (2009) *Ecology without Nature: Rethinking Environmental Aesthetics*. Cambridge, MA: Harvard University Press.

Morton, T. (2013) *Hyperobjects: Philosophy and Ecology after the End of the World*. Minneapolis: University of Minnesota Press.

Morton, T. (2016) *Dark Ecology: For a Logic of Future Coexistence*. New York: Columbia University Press.

Nancy, J. L. (2015) *After Fukushima: The Equivalence of Catastrophes*. New York: Fordham University Press.

Nancy, J. L. and Barrau, A. (2015) *What's These Worlds Coming To?* New York: Fordham University Press.

Neyrat, F. (2019) *The Unconstructable Earth: An Ecology of Separation*. New York: Fordham University Press.

Ostrom, E. (1990) *Governing the Commons: The Evolution of Institutions for Collective Action*. Cambridge: Cambridge University Press.

Ostrom, V. (1997) *The Meaning of Democracy and the Vulnerability of Democracies*. Ann Arbor: University of Michigan Press.

Povinelli, E. (2016) *Geontologies: A Requiem to Late Liberalism*. Durham: Duke University Press.

Purdy, J. (2015) *After Nature: A Politics for the Anthropocene*. Cambridge, MA: Harvard University Press.

Rabinow, P. (2011) Dewey and Foucault: what's the problem? *Foucault Studies* 11: 11–19.

Reghezza-Zitt, M. and Rufat, S. (2019) Disentangling the range of responses to threats, hazards and disasters. Vulnerability, resilience and adaptation in question. *Cybergeo: European Journal of Geography* 916.

Sen, A. (1999) *Development as Freedom*. Oxford: Oxford University Press.

Thacker, E. (2010) *After Life*. Chicago: University of Chicago Press.

Tocci, N. (2019) Resilience and the Role of the European Union in the world. *Contemporary Security Policy*. doi:10.1080/13523260.2019.1640342.

UN-Habitat (2014) *Pro-Poor Urban Climate Resilience in Asia and the Pacific*. Nairobi: UN-Habitat.

Wakefield, S. (2017) Inhabiting the Anthropocene back loop. *Resilience: International Policies, Practices, Discourses* 6(2): 77–94.

Wakefield, S. (2019) *Anthropocene Back Loop: Experimentation in Unsafe Operating Space*. London: Open Humanities Press.

Watts, M. (2015) Now and then: the origins of political ecology and the rebirth of adaptation as a form of thought. In G. Bridge, J. McCarthy and T. Perreault (eds), *The Routledge Handbook of Political Ecology*. London: Routledge, 19–50.

Whyte, K. (2019) Way beyond the lifeboat: an indigenous allegory of climate justice. In K.-K. Bhavnani, J. Foran, P.A. Kurian and D. Munshi (eds), *Climate Futures: Reimagining Global Climate Justice*. London: Zed Books.

2

RESILIENT EARTH

Gaia, geopolitics and the Anthropocene

Simon Dalby

> Resilience is the capacity of a system, be it an individual, a forest, a city or an economy, to deal with change and continue to develop. It is about how humans and nature can use shocks and disturbances like a financial crisis or climate change to spur renewal and innovative thinking.
>
> *(Stockholm Resilience Centre 2015)*

Introduction

Earth System Science has recently picked up many of the themes in the earlier discussions of James Lovelock's Gaia theory. It poses questions of the earth as a self-regulating system, of life as a key player in the earth system, and the possible outcomes of the rapid rise of greenhouse gases in the atmosphere in particular. While these matters have until recently seemed far from the concerns of social theorists, now with the rapid proliferation of discussions of the Anthropocene in particular, and earth system thinking more generally, the questions of what all this means for security, broadly understood, are unavoidable.

Much of resilience thinking has focused on human communities and their abilities to deal with disruptions. State efforts to shape societies in ways that make climate change impacts easier to deal with are in their early stages in many places, but the question of how to make the global system itself more resilient has only been mooted in a few places. As climate change accelerates, its impacts on the global economy and political arrangements seem likely to grow too, posing novel global governance challenges.

Looking to the larger discussion of the earth system science and the planetary boundaries framework this chapter poses the question of how the long-term trajectory of the earth system might better inform social thinking about global security. What novel senses of globality and the dramatic transformations wrought by globalisation might be usefully reconsidered in light of earth system science? How might the nascent discussion of Anthropocene security be enhanced by taking the long-term view of earth history, and scaling up resilience thinking to the scale of global politics?

Earth system science and the Gaia Hypothesis

In the 1970s James Lovelock published a series of papers on a theory of the earth as a self- regulating system, and summarised his ideas in a short book simply called *Gaia: A New Look at Life on Earth* (Lovelock 1979). In retrospect this contribution can be seen as a substantial intervention in the ongoing debate of the period concerning how the world is understood as a matter of environmental politics. It fed into numerous environmental discussions, and theological discussions too. Metaphors of the planet as a sick organism in need of medicine (Lovelock 1991) or forms of holistic therapy were complemented with Gaian atlases of planetary management (Myers 1984/1993). The metaphors of Gaia as a goddess weren't helpful at times, but Lovelock's crucial texts talked not of an intentional presence or external design but instead of the earth as a self-regulating system able to adapt to external change in terms of increasing solar radiation, and moving continents with their related climate impacts over the long term, as well as the short-term consequences of major volcanic episodes and meteor impacts.

Nothing in Lovelock's or subsequent earth system analyses suggests that this context will necessarily be congenial to large-scale organised human life in the long run. The nightmare scenarios relating to the current trajectory of climate change suggest that the planet may indeed become uninhabitable for humans (Wallace-Wells 2017). But clearly both the popular Gaia Hypothesis and the subsequent academic earth system science formulations suggest that at the largest of scales the planetary system, with its self-regulating feedback systems and clear geological record of recovering from catastrophe, and life overall are, in this crucial sense, resilient.

Lovelock's hypothesis suggested that negative feedbacks could function in a way that allowed the planetary system to remain loosely within conditions that favoured life and do so autonomously as a loosely self-regulating system. Life adapts to changing conditions, but in the process also changes those conditions. In the terms used subsequently by the International Geosphere Biosphere Programme 'life is a player, not a spectator' (IGBP 2001: 4) in earth processes. Hence life itself matters as a force of nature. Now, as human activities have expanded in dramatic fashion, especially so in the 'great acceleration' period since World War II (McNeill and Engelke 2016), some of the earth system scientists are thinking in terms of a new component to this system, a rapidly expanding technosphere that is a novel entity in the system and a key to contemporary transformations (Zalasiewicz et al. 2017).

As anthropogenic caused climate change accelerates, it is far from clear that human systems can be adapted rapidly enough to ensure that the form this resilience takes is conducive to large-scale human civilisation. The implications of the earth system analysis do make it clear that decisions about what to make, how to build cities, grow food and power societies, the specifics of how the technosphere is constructed, are key to which pathway humanity takes in coming decades, towards a relatively stable earth or towards a 'hothouse' one with major climate disruptions (Steffen et al. 2018). The discussion about all this, summarised by the invocation of the term Anthropocene to emphasise the rapid rise of humanity as an earth system scale transformation agent, poses the question as to whether human civilisation will be resilient enough to survive the changes it has, until recently mostly inadvertently, set in motion.

At the largest scale, the planetary system may be resilient, and at the smaller scales of individual ecological assemblages, some of these too may be resilient to at least some of the shocks set in motion by industrial civilisation. The question the Anthropocene formulation poses is whether human civilisation has the resilience to deal with its own self-inflicted difficulties – only most obviously nuclear weapons, the extinction crisis and climate change, all artefacts of the most dynamic phase in the history of capitalist expansion that has occurred in the period aptly termed the great acceleration. Climate change is already introducing shocks to the system; at the global scale, resilience now implies the need for rapid decarbonisation. This new context for geopolitics implies a very different set of security priorities for states and citizens; the 2015 Sustainable Development Goals formulate matters in terms of the need for transformation (United Nations 2015).

Geopolitics

Grappling with this formulation foregrounds questions of geopolitics now that it has become clear that the actions of the rich and powerful among us are dramatically altering some crucial parameters of how the earth system operates. While the term geopolitics frequently invokes considerations of the military dimensions of rivalries among empires, over the long term matters of geopolitical economy, quite literally what gets made where with what societal consequences, are much more important in shaping the patterns of global politics (Agnew 2003). These matters are now crucial to shaping the future of the earth system, something that has been slow to be recognised by either the scholarship in the field of international relations or the policy communities that concern themselves with global matters, and the functioning of international institutions and the United Nations system. This novel human context is now key to considerations of the future of geopolitics and who tries to secure what and whom where.

A focus on production, on quite literally the processes that make the future through how the technosphere changes the operation of the earth system, is now key (Dalby 2014). But this is, as yet, seriously out of line with the traditional discussions of geopolitics and the grand strategy of major states whose policy makers still assume a stable geographical context for their rivalries. It is also out of line with most social science discussions of capitalism and social change, many of which don't investigate the material specifics of what industrial systems, either neoliberal or state capitalist, produce and how. Measures of GDP don't capture the key ecological consequences of the mix of goods and services in their calculations (Gopel 2016).

Both the accelerated interconnections of the global economy and the rapidity of environmental change challenge the logics of state security and the assumptions of sovereign states as the basic structure of global politics which provides security in meaningful ways (Stiglitz and Kaldor 2013). Resilience thinking needs to grapple with both the scale and speed of contemporary changes. But if it is to be useful in terms of global politics it needs to be stretched beyond how it has usually been invoked in recent political discussions of either security or environment. The analogies between natural and human systems, and the attempts to couple social and natural changes into analyses of socio-natures, point to the conceptual complexity of the task; the Anthropocene formulation provides an overarching contextualisation (Grove, 2018; Lewis and Maslin 2018).

The scale of contemporary transformations make it clear that environmental protections of particular ecosystems in situ is no longer the appropriate framing of current tasks. The complexity of the earth system and the scale of human activity make it clear that resilience understood in terms of system recovery after a disruption isn't anything like enough to grapple with the overall crisis, however efficacious it may be on smaller scales if updated as hacking or something similar by on the ground activists (Chandler 2018). Neither do state centric formulations of security grapple effectively with the scale of contemporary transformations.

Disaster capitalism has proven effective at taking advantage of crises at the expense of both people and disrupted environments, but there is no indication that this is sustainable at the big scale or over the long term. Indeed Klein's (2007) analysis of the shock doctrine has long suggested precisely that resilience in this sense is currently about the persistence of the worst aspects of predatory economic practices, not about common interests in human or ecological flourishing. Her recent update in terms of the aftermath of Hurricane Maria in Puerto Rico suggests that the Trump administration operates on precisely these premises (Klein 2018).

Climate security

Resilience has become a key theme in security thinking in the last couple of decades as the Anglosphere remilitarised international politics in the aftermath of 9/11 and its war on terror that has turned into a perpetual war in South West Asia. But careful analysis of the genealogy of this term suggests that it was much more about rhetorical innovation than substantial policy change. Maintaining social order in turbulent times is as much about tackling what is seen as contagious disorder as it is about protecting populations (West 2018). Keeping calm and carrying on is after all a slogan invented in the early 1940s in the UK in anticipation of a Nazi invasion. Making subjects resilient in the absence of effective governance in a crisis is also frequently about individual survival rather than community solidarity, although in a crisis, as disaster studies emphasise, communal action is frequently the social response (Solnit 2010). But once the initial disaster has passed frequently economic liabilities render the poor unable to rebuild their lives.

Innovative responses to vulnerabilities in particular places, not least the imaginative use of digital tools that evade state surveillance, suggests possibilities for small scale adaptation that gets beyond state understandings of resilience. But as residents of Canada's capital city discovered once again in September 2018, such innovations are dependent on electrical grids and cellphone towers that are especially vulnerable to storms, or, in this particular case, tornados. Communications infrastructure is key to responses to storm events, and if it isn't robust enough to survive in a crisis, as storms and floods in both New Orleans and Mumbai a decade ago demonstrated, then it offers little by way of security for those in harm's way (Dalby 2009). Imaginative digital hacks to respond in the face of catastrophe are likewise dependent on the network infrastructure and in turn on the corporations that build it.

In so far as security is related to states and sovereignty protection it is organised in a way that emphasises social stability and short-term interests rather than collective or long-term welfare. Indeed what Paul Harris (2013) calls the cancer of Westphalia is antithetical to global climate policies. Invoking global security suggests a larger canvass and the possibilities of cooperative action. Complex interconnected crises are a matter of much greater concern at the global scale as both the financial crisis of 2007–8 and the global food disruptions a couple of years later emphasise (Homer-Dixon et al. 2015). But in so far as this discussion reverts to thinking about how to maintain political stability in the international system in the face of specific instances of environmental change, it emphasises only resilience in the existing political system (Goldstein 2016). The problem here is simply that it is precisely this political system that is perpetuating modes of economy powered by fossil fuels that are causing accelerated climate change. While much of the climate security discussion concerns itself with small scale peripheral conflicts that may or may not be related to climate, what is not in much of this discussion is the larger question of global instabilities, of potential major power wars caused by either climate change, or attempts to adapt to it, or mitigate it by geoengineering (Dalby 2015).

Much of the security discussion related to climate focuses on fragile states and agricultural societies without the resilience capabilities to rapidly adapt to drought (Ide 2018). Focused more specifically on instabilities in particular places, the climate security discussion is usually about South and South West Asia and sub Saharan Africa (Busby 2018). State capacities, and the willingness of elites to consider more than their own narrow short-term interests, are a key factor preventing poverty, famine and distress becoming conflict situations (Mobjork, Smith and Ruttinger 2016). While policy innovations to make these, mostly agricultural regions, more resilient in the face of increasingly unpredictable weather and severe storms and droughts are necessary parts of climate change adaptation, they don't grapple with the key issues of how the global economic system will be shaped in coming decades. Rather than grappling seriously with the wholesale transformations, currently in motion, these versions of climate security perpetuate metropolitan anxieties about peripheral instabilities. To a very large extent they do this because the implicit geopolitical categories that shape policy discourse still rely on long outdated assumptions concerning the planet and its geopolitics (Dalby 2013).

John Agnew's succinct summary of geopolitical reasoning suggests:

> the world is actively 'spatialized,' divided up, labeled, sorted out into a hier-
> archy of places of greater or lesser 'importance' by political geographers, other
> academics and political leaders. This process provides the geographical framing
> within which political elites and mass publics act in the world in pursuit of
> their own identities and interests.
>
> *(Agnew 2003: 3)*

This framing now has to be supplemented by a clear sense of the significance of
non-stationarity, recognition that past geographical patterns of ecological phenomena, only most obviously those of climate, are changing and the geopolitical
reasoning that political leaders use can no longer assume stable geographies.

> [W]e are entering a new unstable, and unpredictable geological era that will
> endure for thousands or tens of thousands of years ... now we must face up to
> the fact that this situation, an irreversible and dangerous trajectory, is our
> future and the ideas that we have inherited from the era before the break must
> all be open to question.
>
> *(Hamilton 2017: 37)*

The political questions for our time concerning global security are about how
the future changes will be shaped by geopolitical reasoning in coming decades.
Assumptions about how interventions in the climate system might be made effective now supplement the 'spatializations' of geopolitical reasoning. They do so
precisely because earth system science makes clear just how dynamic the earth
system is, and that the rich and powerful parts of humanity have become a force of
nature on the scale of geological processes. Nature is no longer the backdrop for
the human drama, but how it is reshaped by investment decisions, land use planning, the construction of dams, irrigation systems and forestry policy are now key
parts of the processes of constructing the geopolitical context for future decades.

Anthropocene

Whether the earth system's resilience turns out to be benevolent or disastrous in
coming decades depends in part on how the politics of production and the energy
systems to power the global economy play out. This too depends on some simple
assumptions about geopolitics, and crucially whether activists, corporate leaders and
politicians operate on the assumptions that there is nothing that can be done about
climate change, and that violent measures to maintain some form of social control
by the continued use of firepower is the only option, or move to take active
measures to transform the world economy, as the sustainable development goals
require (Dalby 2018a). Invoking 'climate terror' with portrayals of imminent disasters as unavoidable, and hence with political imputations that nothing can be

done, have to be tackled to ensure climate is dealt with rather than ignored (Chaturvedi and Doyle 2015). The assumptions that structure much of the rationale for the Trump administration's demolition of such things as fuel efficiency standards is precisely that nothing can be done, and the social costs to the existing economy are too high to try to tackle climate change (Eilperin, Brady and Mooney 2018). The assumption that the next stage of the Anthropocene is inevitably disastrous belies the possibilities of radically tackling current trajectories.

Critics of discussions of the Anthropocene are quick to insist that conflating humanity into one uniform entity obscures the political responsibilities of specific parts of the species for causing climate change in particular, and other disruptions in the period of the great acceleration too (Moore 2016). 'Beneath sweeping generalizations about how "humanity," "modernity," or "we" have caused the Earth's changes during the Holocene, the conflicted complexities of race, ethnicity, and civilization largely are ignored without nuance in most Anthropocene debates' (Luke 2018: 2). The post-political invocation of universal dangers that can only be responded to by market and technical means, 'climate inc' in Paul Wapner and Hilal Elver's (2016) terms, eviscerates the Anthropocene of any clear politics, contestation and a clear antagonist against whom to mobilise. 'Climate terror' results.

There are important exceptions to this, powerfully articulated in specific places by the opposition to many extractivist projects, mines, dams and pipelines in particular, places that Naomi Klein (2014) so succinctly discusses in terms of 'blockadia'. Likewise the struggles for economic control over the future relate to attempts to challenge the fossil fuel companies directly, not just on the ground in particular places but in terms of political interventions to try to facilitate the rapid expansion of renewable energy and undercutting the investment strategies of the fossil fuel companies. Hence the importance of arguments from the divestment movement in particular that a focus on the sources of environmental change, specifically the fossil fuel companies and related state subsidies of their activities, have to be formulated as an enemy that must be mobilised against, a clear antagonist that is profoundly threatening. Related to this formulation fossil fuel critics have invoked metaphors of wartime emergency to justify a rapid mobilisation of industrial capabilities to build renewable energy systems and phase out fossil fuels (Mangat and Dalby 2018).

While the earth will, in the long run, deal with the spike in carbon dioxide that fossil fuelled economies have generated in the period of the great acceleration, the questions surrounding resilience as a political concept enmeshed in these processes relate to the scale of transformations and their rapidity. In its conventional formulation resilience related to climate is mostly about climate adaptation and recovering after disasters in particular places. Resilience also doesn't think much about pre-emption; linked to climate it's mostly about adaptation, and rebuilding after the fact, not necessarily building sensibly in the first place. Mostly the scale in this discussion is at best a matter of nation states. But the formulation quoted in the epigraph of this chapter from the Stockholm Resilience Centre (SRC) suggests that innovation is a key part of their use of the term; it's about more than reconstruction after disruption; it's potentially about more creative changes which, because they look to the global scale as well as specific ecosystems in many places, may evade the generalisations of climate terror and the invocation of an homogenous and hence apolitical humanity.

The Stockholm Resilience Centre

At the global scale the key institution using the concept of resilience is the Stockholm Resilience Centre. Its researchers have been central to the discussion of earth system boundaries, coupled social and natural systems, food production, sustainability and complex systems analysis. As such their focus on resilience requires detailed explication in any attempt to think through how resilience works at the largest of scales. As the epigraph to this chapter suggests, the discussion of resilience in the SRC framework focuses in part on the ability of systems to recover from shocks or disturbances (SRC 2015). The second sentence in the SRC epigraph introduces a normative element, the assumption that change can lead to desirable outcomes by spurring 'renewal and innovative thinking'. The possibility that it might lead to retrograde actions is implied but the focus on complex systems and the coupling of nature and humanity suggests that the possibilities for doing things differently are a key part of the Stockholm formulation. Responding to shocks positively is an aspiration that is key to the Stockholm formulation of resilience, one that critics might be quick to dismiss, but is essential to the Centre's work.

The contrast with warnings about disaster capitalism, most notably in Naomi Klein's (2007) work on the theme, is stark, and it raises key political questions about the potential of resilience understood in Stockholm terms. Crucial to this is the question of who has the resources to respond to disruptions, and how. The politics of this is fundamental, because clearly in response to the financial crisis of 2008 the transformations of the global economy, that it indicated were essential, failed to materialise. This was not least because of a failure to understand the necessity of changing a vulnerable system, and the raw power of decision makers to reboot the failed financial system by what effectively were massive bailouts by public institutions. While this may have allowed the financial system to survive, its continuation depended on state intervention: an implicitly embarrassing admission that its neoliberal ideological supports based on supposed market self-corrections were bankrupt. Nonetheless, in the absence of an effective counter narrative or a politically effective coalition of progressive forces, what happened was a bailout of the finance system, without a reconfiguration to deal with either its internal failings or its ecological unsustainability. This failure to deal with ecological unsustainability has become a key theme in recent Stockholm thinking, focused on the dangers of crossing tipping points in the global system that may lead to a 'hothouse' pathway rather than a pathway to a stable earth (Steffen et al. 2018).

Rejecting modern notions of humanity as separate from ecological processes suggests that understanding natural and human systems as interlinked is essential to resilience thinking. 'This means that in our globalised society, there are virtually no ecosystems that are not shaped by people and no people without the need for ecosystems and the services they provide' (SRC 2015). The failure to take this insight into the human condition seriously lies at the heart of many problems and hence the necessity of reconnecting to the biosphere. Humanity has changed environments in numerous ways to feed a rapidly growing population: 'But the gains achieved by this spectacular re-engineering have come at a price. It is now widely apparent and acknowledged that humanity's use of the biosphere is not sustainable' (SRC 2015). Hence the necessity of rethinking how humanity does things, and doing research to generate knowledge about how to cope with the stresses humanity has generated. What is interesting in the Stockholm framework is that: 'It is about finding ways to deal with unexpected events and crises and identifying sustainable ways for humans to live within the Earth's boundaries' (SRC 2015). This requires linking three key themes together; first, the complex interdependence of people and ecosystems; second the extraordinarily rapid acceleration of change in the last couple of hundred years and since the middle of the twentieth century in particular; and third, the number of innovations opened up by this acceleration which have generated numerous social and technical possibilities for dealing with unsustainability. Modernity's most pernicious dichotomy, of humanity as apart from rather than part of nature, is no longer tenable in such thinking.

Sustainability

Sustainability is defined in the Stockholm framework as living within earth system boundaries, a series of threshold conditions within which earth system science suggests civilisation can thrive. These are conditions loosely analogous with the last 12,000 years of the Holocene, a remarkably stable period in the earth system where humanity has been able to thrive (Rockström et al. 2009; Steffen et al. 2015). Maintaining the earth system in these conditions is, the argument suggests, necessary because the alternative, frequently present in the geological history of prior periods, is of rapid oscillations between glaciations and brief 'interglacials'. Now however the combustion of fossil fuels and the rapid transformation of terrestrial ecosystems, by rapid deforestation and the spread of industrial agriculture, have pushed the earth system into a condition for which there is no recent geological analogy.

Given its remarkable climatic stability in comparison to the rapid fluctuations between glacial and interglacial periods over the previous few hundred millennia, the Holocene is understood as a safe operating space for human civilisation. The fossil fuelled economy of the present has pushed the system well outside the boundaries within which we understand the historical dynamics of the earth system in at least rough outline. Changing the global economy rapidly to move it along a sustainable pathway towards a stable earth will require fundamental changes to society, ones plainly incompatible with contemporary neo-liberal assumptions that markets and growth are essential.

> The Stabilized Earth trajectory requires deliberate management of humanity's relationship with the rest of the Earth System if the world is to avoid crossing a

planetary threshold. We suggest that a deep transformation based on a fundamental reorientation of human values, equity, behavior, institutions, economies, and technologies is required. Even so, the pathway toward Stabilized
Earth will involve considerable changes to the structure and functioning of the
Earth System, suggesting that resilience-building strategies be given much
higher priority than at present in decision making.

(Steffen et al. 2018: 6)

Resilience in terms of the ability to deal with disasters and shocks will be
essential to future human societies, even if a stabilised pathway emerges shortly as
the preferred human strategy. Scaling up innovations in agroecology, transition
towns and such things as ecosystem fisheries arrangements are key to sustainability
in the future. How to use disturbances and crises as a stimulus to innovation is a
key theme that links SRC resilience thinking to the tasks of transformation item-
ised in the sustainable development goals. Key to this, although frequently under-
reported in mainstream media focused on national politics, is the emergence of
numerous initiatives by municipal actors, corporations and other non-state entities
to tackle climate change, especially where national governments explicitly deny
climate is a policy priority (Bernstein and Hoffman 2018).

Critics of the Anthropocene formulation frequently focus on such terms as
'deliberate management' to suggest that earth system boundaries, safe operating
spaces and the related terms in the Stockholm formulation feed into a politics of
authoritarian global managerialism in the guise of some form of cosmopolitanism
(Moore 2016; Chandler, Cudworth and Hobden 2018). From there it is an easy
step to point to the potential for political elites to try solar radiation management
and other technical fixes to attempt to maintain the profoundly unjust social system
of the present while trying to ride out the consequences of its production systems.
Taking earth system science seriously suggests that there are no single points of
intervention in the system that will allow for such control. The key ontological
shift from modernity with its assumptions of humanity as separate from nature, to
an earth system understanding of the rich and powerful parts of humanity as
interlocked in system-influencing activities, requires much more than technological
attempts to intervene in only a few crucial dimensions.

The Stockholm authors are clear that a deep transformation of many things is
going to be needed, well beyond elite fiddling with technology while the planet
burns. Part of what is being suggested as necessary is rapid technological innovation
(Rockström and Klum 2015) and this too is easily interpreted as the whole answer.
If this is the case then critics are right to argue that this is unlikely to be enough. As
Hamilton (2017) suggests in his arguments about the Anthropocene, this is likely to
lead to a very bad Anthropocene for much of humanity rendered increasingly
vulnerable by climate change in particular. Crucial to the recent Stockholm formulations is a necessity to mobilise innovative financial measures to rapidly build
more sustainable modes of economy. What is missing in much of the Stockholm
discussion is a political strategy to do just this. But as Bernstein and Hoffman (2018)

emphasise, social and political innovations are underway in many venues; the opposition to the Trump administration suggesting that political authority is increasingly contested as the dangers of its attacks on nascent global governance arrangements are increasingly recognised.

Financing sustainability?

The necessity of doing finance in particular, and much of industrial production, very differently is clear both in terms of the intellectual inadequacies of traditional economics (Gopel 2016) and the need to radically rethink conventional economics to take earth systems seriously and develop modes of living that function within the safe operating space (Raworth 2017). That is, it is necessary if more than the profitability of the existing extractivist economy is taken into account, a matter that simply isn't part of the calculus of political elites in most parts of the world, at least not yet. While this may be beginning to change, and the universal adoption of the sustainable development goals (United Nations 2015) and subsequently the Paris Agreement on Climate Change in 2015 (Falkner 2016) indicate at least rhetorical recognition that carboniferous capitalism is unsustainable, clearly much more needs to be done in terms of making financial institutions, and not just the banking sector, not only recognise the risks from climate change but to change their approval criterial to actively build ecological sustainability into their procedures. In terms of resilience being effective responses to crises, the losses from climate related damage are rapidly mounting, but whether this will foment the necessary rethinking of how finance works remains to be seen. Which is why the politics of the divestment movement is important in so far as it focuses attention on investment decisions that have long-term ecological consequences and suggests a crucial venue in which activist politics can supplement the protest spaces in campaigns against extractivism (Routledge 2017) and engage with major corporations and the financial sector in ways that shape production decisions (Mangat, Dalby and Paterson 2018).

The potential to shape future economies, so they pay attention to planetary boundaries and incorporate these ideas of sustainability into their planning, may be there, as the Stockholm authors argue, but how the politics of this will play out needs urgent attention. It does so not least because the climate denial tendencies in the Anglosphere and the dramatic reassertion of what Cara Daggett (2018) calls petro-masculinities operate to delay and obstruct much needed innovations in international institutions while perpetuating the fossil fuel production that is at the heart of the climate change problem. Thus this geopolitical imaginary of firepower as the key to the future, denying the ecological consequences of its actions, or if these are conceded (Dalby 2018a), denying that there is much that can be done about them, stands in stark contrast to what the Stockholm Resilience Centre advocates. The recent rise of nationalist, if not xenophobic, political movements has made everything more difficult, although simultaneously more fragile (Hedges 2018). In particular, the insistence on sovereignty, wall building and rhetoric that pose migrants as threats to national communities operates to further obscure the ecological and economic links that matter across borders (Dalby 2018b). This isn't resilience, it's domination by a financial and state system that relies on force to perpetuate its expansion despite the disruptions that inevitably follow (Sassen 2014).

With notable exceptions, like Mark Carney, governor of the Bank of England, and the United Nations project on sustainable finance (United Nations Environment Programme 2018), international finance has not taken either the vulnerabilities to climate risks or the transformation of ecosystems, hydrological fluctuations or biodiversity loss into consideration in its calculations. Addressing these failures in terms of 'Sleeping Financial Giants', once again the SRC has focused on finance and investment as key to climate stability (Gaffney et al. 2018). However in terms of resilience, and the framework of using shocks to stimulate appropriate behaviours, the prefatory quote to the report by Mark Carney is especially poignant: 'Once climate change becomes a clear and present danger to financial stability it may already be too late to stabilize the atmosphere at two degrees' (cited in Gaffney et al. 2018: 3). In these terms current human elites may not be resilient given their failure to adapt and innovate in time to deal with climate change.

While their critics might celebrate such a demise of the financial class, it will probably be too late to prevent at least some of the serious disruptions of the hot-house pathway. The urgency of dealing with the potential tipping points of boreal forests and the destruction of the Amazonian tropical rain forest is outlined in detail in Sleeping Financial Giants because these are two of the earth system components that might well tip the overall system into a new hothouse pathway where human actions may be unable to arrest rapid disruptive change caused by global heating and positive feedbacks, like the melting of the Arctic ocean ice cover. If as current trends suggest, boreal forests and tropical rainforests start to burn more extensively due to drying and heating, and then become sources of carbon dioxide rather than sinks absorbing it from the atmosphere, a climate tipping point will have been crossed, and rapid destabilisation of many other parts of the system is the likely result.

Politics and the Anthropocene

The nightmare scenario looming in the near future is precisely that this will happen and political elites will respond by trying to artificially lower the amount of solar radiation reaching the earth's surface in a variety of geoengineering experiments with unknowable consequences both for the earth system and global politics (Dalby 2015). Avoiding this future is key for sustainability, and the clear statements from the SRC and its scholarly network suggest that this requires adaptations in numerous places, not only in how the finance sector considers what to invest in and fund. The implications of the earth system analysis are that technical solutions imposed on the system in attempts to engineer a future are unlikely to work. Much more comprehensive solutions in numerous parts of construction, agriculture and infrastructure are needed to simultaneously buttress local environments against disruptions while simultaneously working to enhance the functioning of the earth system at the global scale. In James Lovelock's (2014) terms failure to deal effectively with at least the worst disruptions of the earth system set in motion by fossil fuelled economic systems seem likely to make it a very 'rough ride to the future', a 'hothouse earth' in SRC terms.

What is not clear is that the term resilience as currently used actually has the political gravitas or the obvious resonance to invoke these widespread changes in numerous locales simultaneously. It is noteworthy that the sustainable development goals explicitly formulate their task in terms of transformation, not resilience. The SRC authors understand the need for rapid transformation, but caution both that new investments and a more sustainable economy will also have to build in resilience in the sense of being able to adapt to shocks given how much change in the earth system has already been set in motion, and 'Yet time is short – financial actors, and humanity at large, need to wake up and recognize the new and urgent challenges posed by nonlinear dynamics in the Earth System' (Gaffney et al. 2018: 7).

But if the earth system is to be resilient in the sense of adapting within the parameters that have allowed human civilisation to flourish, rather than, as Lovelock's pessimistic commentaries have sometimes suggested, disposing of humanity altogether, then rapid innovation will be needed in many parts of the global economy to shape the technosphere in directions that make it more flexible to adapt to the novel configuration of the earth system. There is of course also the old argument, also reiterated by James Lovelock (2014) in his more recent ruminations on Gaia, which suggests that humanity may merely be a stage on the way to the emergence of a global mind, an electronic intelligence that might yet be Gaia become conscious and capable of deliberate self-regulation. Perhaps this might be understood as the triumph of a self-conscious technosphere? As to whether humanity is a useful part of such a world or a dangerously destabilising element is still probably a question best tackled by science fiction authors rather than social scientists.

The Stockholm formulations suggest a much more ambitious global agenda than that encompassed by popular understandings of resilience as fortitude and adaptability in the face of unavoidable vicissitudes. They also suggest much more than resilience as a strategy for containing political contagion. While the SRC has been key to developing the research and thinking that both formulated the earth system boundaries and the ideas of safe operating space, ironically their ideas of hothouse pathways or a stable earth, of innovation and coupled human and natural systems, have much more purchase on the political tasks of the present than the term resilience usually implies.

Now the key questions for social scientists interested in shaping the agenda for the next stage of the Anthropocene are how to formulate politically efficacious strategies to move ahead with social and technological innovations in particular contexts, while simultaneously accelerating the phase out of fossil fuelled modes of human life and recovering political control of financial instruments from the speculative financialisation that dominates the contemporary global economy. In so far as these transformations succeed, human civilisation in general may in future deserve the moniker resilient.

References

Agnew, J. (2003) *Geopolitics: Re-Visioning World Politics*. London: Routledge.

Bernstein, S. and Hoffmann, M. (2018) The politics of decarbonization and the catalytic impact of subnational climate experiments. *Policy Sciences* 51(2): 189–211.

Busby, J. (2018) Taking stock: the field of climate and security. *Current Climate Change Reports* 4(4): 338–346.

Chandler, D. (2018) *Ontopolitics in the Anthropocene: An Introduction to Mapping, Sensing and Hacking*. London: Routledge.

Chandler, D., Cudworth, E. and Hobden, S. (2018) Anthropocene, Capitalocene, and liberal cosmopolitan IR: a response to Burke et al.'s 'Planet Politics'. *Millennium: Journal of International Studies* 46(2): 190–208.

Chaturvedi, S. and Doyle, T. (2015) *Climate Terror: A Critical Geopolitics of Climate Change*. London: Palgrave.

Daggett, C. (2018) Petro-masculinity: fossil fuels and authoritarian desire. *Millennium: Journal of International Studies*. doi:10.177/030582981875817

Dalby, S. (2009) *Security and Environmental Change*. Cambridge: Polity.

Dalby, S. (2013) The geopolitics of climate change. *Political Geography* 37: 38–47.

Dalby, S. (2014) Rethinking geopolitics: climate security in the Anthropocene. *Global Policy* 5(1): 1–9.

Dalby, S. (2015) Geoengineering: the next era of geopolitics? *Geography Compass* 9(4): 190–201.

Dalby, S. (2018a) Firepower: geopolitical cultures in the Anthropocene. *Geopolitics* 23(3): 718–742.

Dalby, S. (2018b) Geopolitics in the Anthropocene. In A. Bergeson and C. Suter (eds), *The Return of Geopolitics*. Zurich: Lit, 149–166.

Eilperin, J., Brady, D. and Mooney, C. (2018) Trump administration sees a 7-degree rise in global temperatures by 2100. *Washington Post*, 28 September.

Falkner, R. (2016) The Paris Agreement and the new logic of international climate politics. *International Affairs* 92(5): 1107–1125.

Gaffney, O., Crona, B., Dauriach, A. and Galaz, V., (2018) *Sleeping Financial Giants: Opportunities in Financial Leadership for Climate Stability*. Stockholm: Stockholm Resilience Centre.

Goldstein, J. (2016) Climate change as a global security issue. *Journal of Global Security Studies* 1: 95–98.

Gopel, M. (2016) *The Great Mindshift: How a New Economic Paradigm and Sustainability Transformations Go Hand in Hand*. Heidelberg, Germany: Springer Nature.

Grove, K. (2018) *Resilience*. New York: Routledge.

Hamilton, C. (2017) *Defiant Earth: The Fate of Humans in the Anthropocene*. Cambridge: Polity.

Harris, P. G. (2013) *What's Wrong with Climate Change and How to Fix It*. Cambridge: Polity.

Hedges, C. (2018) *America: The Farewell Tour*. Toronto: Knopf.

Homer-Dixon, T.*et al*. (2015) Synchronous failure: the emerging causal architecture of global crisis. *Ecology and Society* 20(3): 6.

Ide, T. (2018) Climate war in the Middle East? Drought, the Syrian civil war and the state of climate-conflict research. *Current Climate Change Reports*. https://doi.org/10.1007/s40641-018-0115-0

IGBP (2001) International geosphere biosphere programme, global change and the earth system: a planet under pressure. *IGBP Science*, 4.

Klein, N. (2007) *The Shock Doctrine*. Toronto: Knopf.

Klein, N. (2014) *This Changes Everything: Capitalism vs. The Climate*. Toronto: Knopf.

Klein, N. (2018) *The Battle for Paradise: Puerto Rico takes on Disaster Capitalism*. Chicago: Haymarket.

Lewis, S. and Maslin, M. (2018) *The Human Planet: How We Created the Anthropocene*. London: Pelican Books.

Lovelock, J. E. (1979) *Gaia: A New Look at Life on Earth*. Oxford: Oxford University Press.

Lovelock, J. E. (1991) *Gaia: The Practical Science of Planetary Medicine*. Oxford: Oxford University Press.

Lovelock, J. E. (2014) *A Rough Ride to the Future*. Allen Lane: London.

Luke, T. (2018) Tracing race, ethnicity and civilization in Anthropocene. *Environment and Planning: Society and Space*. doi:10.1177/0263775818798030

Mangat, R. and Dalby, S. (2018) Climate and wartalk: metaphors, imagination, transformation. *Elementa: Science of the Anthropocene* 6(1): 58.

Mangat, R., Dalby, S. and Paterson, M. (2018) Divestment discourse: war, justice, morality and money' *Environmental Politics* 27(2): 187–208.

McNeill, J. R. and Engelke, P. (2016) *The Great Acceleration: An Environmental History of the Anthropocene since 1945*. Cambridge, MA: Harvard University Press.

Mobjork, M., Smith, D. and Ruttinger, L. (2016) *Towards a Global Resilience Agenda: Action on Climate Fragility Risks*. The Hague: Clingendael – the Netherlands Institute for International Relations.

Moore, J. (ed.) (2016) *Anthropocene or Capitalocene? Nature History, and the Crisis of Capitalism*. Oakland: PM Press.

Myers, N. (ed.) (1984/1993) *Gaia: An Atlas of Planetary Management*. New York: Doubleday.

Raworth, K. (2017) *Doughnut Economics: Seven Ways to Think Like a 21st-Century Economist*. London: Random House.

Rockström, J.*et al.* (2009) Planetary boundaries: exploring the safe operating space for humanity. *Ecology and Society*, 14(2): 32.

Rockström, J. and Klum, M. (2015) *Big World Small Planet*. New Haven, CT: Yale University Press.

Routledge, P. (2017) *Space Invaders*. London: Pluto.

Sassen, S. (2014) *Expulsions: Brutality and Complexity in the Global Economy*. Cambridge MA: Harvard University Press.

Solnit, R. (2010) *A Paradise Built in Hell: The Extraordinary Communities that Arise in Disaster*. New York: Penguin.

Steffen, W.*et al.* (2015) Planetary boundaries: guiding human development on a changing planet. *Science* 347(6223): 1259855.

Steffen, W.*et al.* (2018) Trajectories of the earth system in the Anthropocene. *Proceedings of the National Academy of Sciences*. Available at www.pnas.org/cgi/doi/10.1073/pnas.1810141115

Stiglitz, J. E. and Kaldor, M. (eds) (2013) *The Quest for Security: Protection Without Protectionism and the Challenge for Global Governance*. New York: Columbia University Press.

Stockholm Resilience Centre[SRC] (2015) What is resilience? Available at http://www.stockholmresilience.org/research/research-news/2015-02-19-what-is-resilience.html

United Nations (2015) *Transforming Our World: The 2030 Agenda for Sustainable Development (A/RES/70/1)*. Available at https://sustainabledevelopment.un.org/post2015/transformingourworld/publication

United Nations Environment Programme (2018) *Making Waves: Aligning the Financial System with Sustainable Development*. Available at http://unepinquiry.org/making-waves/

Wallace-Wells, D. (2017) The uninhabitable earth. *New York Magazine*, 9 July.

Wapner, P. and Elver, H. (eds) (2016) *Reimagining Climate Change*. Abingdon: Routledge.

West, J. (2018) *Defence in Depth: An Anatomy of Containment from Quarantine to Resilience*. Wilfrid Laurier University. Unpublished PhD dissertation.

Zalasiewicz, J.*et al.* (2017) Scale and diversity of the physical technosphere: a geological perspective. *The Anthropocene Review* 4(1): 9–22.

3

SECURITY FOR A FRAGMENTED WORLD

Ecology and the challenge of the Anthropocene

Madeleine Fagan

Introduction

It is becoming increasingly accepted that humans have become a geological force, a development significant enough that the proposal of a new geological epoch – the Anthropocene – has gained currency. Popularized by Crutzen and Stoermer in 2000 (Crutzen and Stoermer 2000), Crutzen (2002) defines it as 'a new geologic epoch in which mankind has emerged as a globally significant – and potentially intelligent – force capable of reshaping the face of the planet'. The insertion of the human into the geological means, argue Crutzen and Stoermer, that there is no 'natural' nature anymore; humans have impacted on it everywhere (Rudy and White 2013: 128). The advent of the Anthropocene then unsettles one of the key organising logics of modernity on which much traditional security thinking is built: the separation between human and nature (Dalby 2009; Dobson 2006; Latour 2004; Walker 2006).

In putting into question the human/nature distinction the Anthropocene poses an important challenge to dominant ways in which the concept of security has been approached in relation to environmental concerns. Security studies literature on the environment and climate change focuses in large part on the question of the implications of climate change for existing questions and theorisations of security; its impact on violent conflict (Kaplan 1994; Homer Dixon 1991; Klare 2001), on human security (Barnett 2001; McDonald 2012); on whether it should be considered a security issue (Deudney 1990); on securitisation theory (Floyd 2010; McDonald 2012); and on how it alters security discourse (Corry 2014; Trombetta 2008). These approaches focus either on what we need to do to our security thinking to better understand, mitigate, manage or map environmental problems, or on the impact of adding environmental concerns to the security agenda. However, the concept of the Anthropocene destabilises the organising categories that animate much of this literature: the distinction between referent objects, the logics of inclusion and exclusion, the idea of agency and a unified human subject, and not least the imagination of an intelligible world as a whole (Colebrook 2012; Dalby 2009; Dobson 2006; Mitchell 2014; Walker 2006).

Ecological security approaches have attempted to address these challenges posed to environmental security thinking in light of the Anthropocene by drawing on concepts from ecology in order to emphasise the interconnected nature of the human and natural, reframe security in terms of *resilience* rather than protection, and to make the referent object the biosphere rather than the human or the environment (Barnett 2001; Cudworth and Hobden 2011; Dalby 2009; Litfin 1999; McDonald 2013; Pirages 2013; Plumwood 2002; Von Lucka, Wellmann and Dietz 2014). However, this chapter argues that there are important limits to the extent to which the pairing of ecology and security is sufficient to the conceptual challenges posed by the Anthropocene.

By analysing ecological approaches to security as a potential avenue for rethinking security in the context of the Anthropocene I will demonstrate the continued dominance and centrality of the nature/culture binary for conceptualising both ecology *and* security, and argue that because of this commonality the scope for a critical reorientation of security thinking from an ecological perspective is limited. I identify some of the assumptions on which 'ecology' rests in order to map how they already inhabit the concept of 'security' such that any attempt to use them as a basis for a critique or reorientation of security thinking ultimately encounters limits. These are assumptions about the human, nature and the world, which Anthropocene thinking puts into question. Instead, I argue that the pairing of ecology and security results in a reproduction of the anthropomorphic assumptions which have arguably contributed to the creation of the Anthropocene epoch whose consequences ecological security is intended to mitigate.

The chapter is arranged in three sections. Firstly, I outline some key features of ecological security approaches and associated accounts of resilience. Secondly, drawing on R. B. J. Walker's concept of the politics of escape, I suggest that in attempting to escape the nature/culture binary the move to ecology in fact simultaneously re-inscribes and obscures this distinction. Finally, I offer a framework for understanding these difficulties as symptomatic of the continued dominance of a fiction of 'the world' in both security and ecological thought and the account of the modern subject that this produces.

Ecological security, resilience and the 'nature' problem

There are a number of key themes drawn from ecological thought which we can see taken up with different importance and emphases in ecological security literature. The first is an intrinsic valuing of the non-human sphere (Eckersley 2007: 305; Plumwood 2002: 9). The value of the non-human develops out of a broader concern with the nature/culture binary, which ecological thought is concerned to disrupt (Cudworth and Hobden 2011: 48; Plumwood 2002: 4). Ecological thought diagnoses this binary as having led to the ecological crises in which we now find ourselves, in part because it has enabled and sustained a worldview in which humans can exert mastery over nature (Plumwood 2002: 10). In contrast, an ecological approach offers an appreciation of human embeddedness in a broader ecological context (Barnett 2001; Pirages 2013). Drawing on earth-systems sciences, the planetary interconnectivity of human and non-human life – or human and 'natural' phenomena – is often conceived of in terms of the 'biosphere' (Barnett 2001; Cudworth and Hobden 2011).

Approaches to security that draw on ecological thought then seem appealing in response to the Anthropocene reframing of the human/nature relationship. They offer a focus on the close ties between the human and non-human world, tracing the implications of an understanding of the world in terms of the complex interdependence of ecosystems (Barnett 2001; Cudworth and Hobden 2011; Dalby 2009; Litfin 1999; McDonald 2013; Pirages 2013; Plumwood 2002; Von Lucka, Wellmann and Dietz 2014). The key move made by ecological security approaches is to reformulate security with the planet, biosphere, ecosystem or ecological processes as its referent object (Barnett 2001; Cudworth and Hobden 2011; McDonald 2013). More specifically, the security of the biosphere is figured in terms of maintaining 'dynamic equilibrium' in the face of challenges posed by political, economic and social structures (McDonald 2013). The notion of equilibrium is developed in terms of 'balance' (Pirages 2013: 143), 'harmony' and 'stability' (Barnett 2001: 109) with respect to the relationship 'between humans and the ecosystems in which they are embedded' (Pirages 2013: 143). As Barnett argues, 'security from an ecological theory perspective therefore involves thinking about the whole rather than the parts' (Barnett 2001: 111).

While 'traditional' security thought in relation to the environment arguably remains bound up in a logic of threat, protection and stasis (Dalby 2009) which cannot respond to the rapid changes likely to characterise the Anthropocene, ecological security puts forward an alternative model of security in terms of resilience (Barnett 2001; Corry 2014). This is an intuitively appealing alternative because the concept of resilience as used here draws on ecological theory, which, in Barnett's (2001: 111) description, refers to a system that can 'retain its organisational structure following perturbation'. Rather than avoiding change, resilience focuses on reducing vulnerability to change (Barnett 2001: 112). Resilience in the ecological context is figured in terms of stability and coherence, the co-evolution of human and natural systems (Corry 2014), and adaptability in order to be able to cope with environmental pressures (Barnett 2001: 134).

Ecological security then looks promising as a route to reframing security for the Anthropocene. However, there is a tension in attempting to bring the 'deep ecological' suspicion of the human/nature dualism into security discourse. The social and natural in ecological security approaches remain separate and distinct, albeit 'co-constitutive, overlapping and intersected' (Cudworth and Hobden 2011: 53). For McDonald (2013: 48), ecological security is about the need to 'fundamentally rebalance the relationship between people and the natural environment'. Barnett's (2001: 115) formulation of ecology is concerned with 'habitat' – the planet (nature) conceived as providing a home for humans. The biosphere is conceptualised here as a more sophisticated mapping of humans-in-nature or humans-in-environment (Pirages 2013: 140).

Recourse to a focus on relations *between* human and nature then potentially obscures some of the more far-reaching implications of ecological political thought. As R. B. J. Walker has argued, to assume any sort of relationship between human and nature is already to reproduce them as separate (Walker 2006: 190), so the ecological security move to make the relationship between human and nature central to security thinking in order to protect them both as part of a broader whole reproduces the human/nature distinction of which 'deep' ecological approaches are suspicious, and which the Anthropocene puts into question. This tension in ecological security discourse is not easily rectified. In fact, it points towards a very difficult problem with attempts to overcome the human/nature distinction in security thinking: that the concept of security itself is so deeply reliant on the human/nature distinction that it cannot be repurposed in service of protecting both nature and the human (see Dobson 2006; Walker 2006).

As Walker has argued, the possibility of modern political life depends on a rupture between human and nature (Walker 2006: 190; see also Dobson 2006; Latour 2004), that is, a particular account of the relationship between them. A crucial part of what makes us modern is the assumption of the separation of the human from an external nature which we can both seek protection from and seek to protect (Walker 2006: 190). The modern subject, on this account, is constituted precisely by its being not-natural, by its separation from the objective world of nature. The primary subject of security discourse has been precisely the securing of this subject as 'modern and properly human, rather than "merely" natural' (Walker 2006: 200). Security discourse then has been founded on a separation between human and non-human – those closer or further away from nature (Walker 2006: 200).

If the modern subject as free, rational, sovereign and autonomous is to be maintained, then securing the human/nature distinction is as much as security can do for nature. Any further acknowledgement of the way in which our conception of the human depends on that which it excludes sets in motion an unravelling of the modern subject that threatens to undermine the edifice of politics, community and security itself. We cannot secure both nature and the human; as Walker argues, 'it will not be enough to think that we can protect nature without also placing our capacity to protect the modern subject into deepest suspicion' (Walker 2006: 200).

As the above discussion has indicated, important aspects of ecological security are very different from neoliberal ideas of autonomy, self-governance, international efficiency and ecological modernisation. But while we might be able to wrestle the concept of security free from a specific neoliberal set of foundations (Corry 2014), it seems more difficult to free it from the broader modern project with which it is so closely entangled, which has arguably caused the ecological crisis to which it seeks to respond, and which is potentially undermined by Anthropocene thinking. If, as

Walker argues, security is about securing the modern subject, then the very suggestion that 'we' can rethink security outside of a modern framework becomes problematic: what would a security look like whose subject was not the modern subject?

Ecology as escape

The ecological critique relies on the idea that to move beyond our current crisis we need to depart from a dualistic understanding of the interior, active, progressive, differentiated, plural, morally countable human (subject) and the passive, singular, static externality of nature (object). The move posited is to interconnectivity. However, it is this approach to ecological thought which allows for the shift back to an ultimate focus on the security of the modern subject discussed above. To claim ecology as an 'alternative' (Barnett 2001: 120; McDonald 2013: 48) to the nature/culture dualism and its attendant violences then begins to look problematic.

Attempting to escape a dualistic framework though a focus on relationship is problematic partly because, as outlined above, to think in terms of relationship is to re-inscribe two separate elements. But more importantly, to posit a focus on the relation between nature and culture as an *alternative* to the human/nature binary is to underestimate the complexities of the operation of that binary in thinking about security. It is not as simple as suggesting that an appreciation of the relationship between human and nature has been absent from security thinking, such that focusing on it allows us to rethink security. One particular account of the relationship between human and nature is already central to security thinking. The demarcation, exclusion and rendering insecure of that aligned with or closer to nature has been a central feature of security discourse. The human/nature dualism relies already on a series of assumptions about the interconnections of human and nature, and on the resulting exclusionary accounts of that relationship. The possibilities and limitations of the modern subject are, as Walker has shown, determined by its *relationship* to a sphere of nature rendered 'outside' (2006: 191). Many dominant frameworks for thinking security are premised already on precisely this relational account of inclusion and exclusion.

A focus on interconnectivity might be read as an acknowledgement of this difficulty and an attempt to offer a different *type* of relationship: an account of the relationship between human and nature which is developed in the context of the planet or biosphere. The ecological move then is to attempt to displace the human/nature opposition by showing how both terms are part of a larger whole, through moving to the planet or biosphere as referent object. However, attempting to escape the human/nature dualism is more difficult than this move to a third term might suggest, and, perhaps more importantly, as I will revisit below, is not necessarily a desirable end. To show this, I draw on Walker's argument about the politics of escape.

Walker (2010) has argued that the desire to escape the system of exclusions central to the politics of the state system is manifested in claims about the need to progress from a politics of the international to a politics of the cosmopolis, globe or planet. This move entails a commitment to projecting the model of progress, security and moral community that is possible within the state, beyond the state (Walker 1997: 74). To escape the problems of the state system is, on this account, to move from fragmentation to integration (Walker 1997). Walker's critique of this equation of escape and integration shows how rather than offering an alternative to the inscription of limits at the boundary of states it simply extends these boundaries outwards (Walker 1997: 75). This does not extend imagination beyond already existing accounts of inclusion and exclusion, but attempts to extend inclusion, transform states into the world, and move from particularity to universality.

Such a move is problematic because, as Walker (1997: 74) shows, the modern states system already offers an account of the relationship between particularity and universality. To then attempt to map a path from one to the other and claim it as progress, and especially to claim universality as an *alternative*, is to misrecognise the great achievement of that system, which is to enable simultaneous claims to both universality and particularity. The modern state system offers a resolution of competing claims to particularity and universality, which is precisely why attempts to articulate progress in terms of moving from the particular or plural to the singular or universal are so limited: that ground for critique is already occupied by the system. I suggest that this pattern operates also in the attempt to transcend the human/nature binary through the ecological move of shifting focus to the planet or biosphere, an attempt which is also articulated in terms of escaping dualism by moving to interconnectivity.

We can see the ways in which the inside/outside framing maps on to the human/nature binary in the claims made for ecological security, above. The ecological move assumes that to think progressively about the environment what is required is a shift away from an understanding of the world as fragmented by the partitioning of human and nature to an understanding of the world as comprising a single, universal, system which integrates human and nature: the biosphere. What is to be protected in ecological security is the wholeness of the whole, the concept of the planet, of the earth as homeland or habitat. It is only within this whole that the concepts of balance, equilibrium, co-evolution and harmonious coexistence can have meaning. Progress is conceptualized as extending the realm of security concern from the human to the biosphere. The move posited is one from the claim that humans are separate from nature, special or different (fragmentation) to the claim that they are part of a larger whole *with* nature (integration).

However, to claim this move as progressive is to misrecognise the way in which the human/nature binary already depends on an account of both a prior whole and a secondary distinction between human and nature. Work is done by the binary to separate human and nature precisely because it relies on a prior assumption that they are not necessarily separate. The move to integration then does not expand or escape the logics already contained within the human/nature binary which already maps out an account of both fragmentation and integration such that shifting between the two still operates within its terms.

Ecological security then becomes entangled in a tension relating to its treatment of singularity and plurality. Interconnectivity offers an attempt to acknowledge the differentiation between human and natural systems while also seeking to dismantle the hierarchical opposition between them. In contrast to an imagination of the human as originally part of the whole of nature – a prior singularity from which plurality is derived – interconnectivity relies on an account of the planet or biosphere in terms of discrete elements which are then gathered together: a prior plurality which forms a singularity.

As the work of Jean-Luc Nancy (1996) has shown, these two accounts mirror one another in their treatment of relation, in that both rely on a first place being given to *either* singularity or plurality. The first – deep ecological – account seeks to move beyond nature as constitutive outside by projecting the subject *into* that outside. The second – interconnectivity – seeks to bring that outside within the logic of subjects-in-relation by imagining it as a product *of* that relation. Both of these approaches try to escape from the human/nature dualism by subsuming one side into the other, offering an account of originary singularity or originary plurality.

However, equating progress with a move to the primacy of either singularity or plurality is, as Walker has illustrated, problematic because it does not move beyond existing accounts of inclusion/exclusion; on this account, the biosphere is an expanded realm of inclusion. This can be demonstrated by the way it is conceived as a realm of coherence, harmony and equilibrium: such features are characteristic of the modern imagination of the human 'inside' of which it is a product. Perhaps more importantly the prominence of such features is also indicative of the difficulties of escaping an anthropocentric logic in which the survival of the human species is the context for understanding the natural world. The Anthropocene, however, as discussed above, confronts us with the unsettling notion that harmony, equilibrium and coherence are no longer (if they ever were) appropriate ways to characterise planetary systems (Hamilton 2015). Moreover, the Anthropocene places into question the role of the human as organising logic around which the world might make sense. The biosphere then seems insufficient to the Anthropocene challenge to anthropocentric modes of organising thought.

The problems encountered in formulating an ecological approach which succeeds in escaping the human/nature opposition are symptomatic of a broader framework within which ecological thought operates. This difficulty is an indication of the way in which such thinking is bound up already in pervasive security logics. The modern world, Walker argues, has been shaped by desires to 'move beyond', but, he continues, there can be no simple move beyond a structuring of 'here and beyond' (Walker 2010: 5–6). While Walker's argument is a critique of claims that we need to move from a politics of the international to a politics of the world, the structure of his argument offers useful insights with which to analyse claims made in ecological security approaches. It is not only that the attempt to move beyond the human/nature distinction ultimately fails, but that the very *desire* to move beyond is itself bound up in the logics which it attempts to exceed.

The ecological critique of liberal modernity as detached and ecologically unsustainable leads to an attempt to move outside of its organising logic. The ecological criticism that we 'misrecognize' (Plumwood 2002) human embeddedness in broader natural systems points to a solution in terms of 'some alternative outside, some world that is somehow more natural, more authentic, more real' (Walker 2006: 192). In this case, a more authentic rendering of the human and its place in the biosphere. But this move outside is, Walker argues, a very difficult task for the modern subject 'who certainly knows well enough the outside he has himself produced but also too easily assumes that to escape to this outside is to escape himself, his own limits, his own responsibilities' (2006: 199).

Escaping to nature, or to claims about the cosmopolitan world, clearly fall into this trap, but so does escaping to the singular interconnected biosphere. The aim of reconceptualising humans as *of* the biosphere is to escape the constitution of the modern subject as one who subjugates nature; to escape the limits of that modern subjectivity by recognising instead the true (natural, authentic, real) nature of being human. We know, Walker continues, how to travel from the domestic to the wild, but not how to think differently about the 'domesticated opposition between the domestic and the wild' (Walker 2006: 199). This domesticated opposition is what is at play in the claims about singularity and plurality highlighted above; the shift to interconnection moves between the two registers but does not escape their opposition.

Ultimately, an ecological approach cannot escape the logic of 'here and beyond', because in seeking the beyond it effectively relies upon the recreation of the world, or planet, as a whole; the recreation of a 'here'. Ecology offers a logic of the cosmos and an account of the modern subject's place within it. It offers an account of the world which determines once again the proper place of the human, albeit it in a more embedded way. It is, as eco-*logy*, bound up in those patterns of knowledge which it wants to critique. As Claire Colebrook argues, ecology is the logic of the earth considered a homeland, 'the epitome of all logics, all systems that claim to express the truth of the world they order' (Colebrook 2012: 199). Whether as part of nature, the planet or the biosphere, ecology offers an account of the proper place of the modern subject; it relies, Colebrook continues, on 'the notion that we might overcome our *narrow* boundedness and self-enclosure and once more find our place in the unified cosmos' (Colebrook 2012: 198). An eco-logy is an attempt at unification and making sense, and so it cannot escape the structuring logics discussed above. This is also why ecology is insufficient to the Anthropocene challenge to the world understood as an intelligible whole. Ecology offers a re-ordering of the world, a recreation of the world as a whole, a neutralising of the threat to logic and sense posed by the Anthropocene.

This matters for thinking about security because to give the modern subject a home is to secure it; it is to reproduce the claims about universality and particularity which constitute the modern subject. Ecological security cannot offer an alternative to the account of security whose subject is the securing of the modern subject. Discourses of ecological or planetary interconnected unity are predicated on the illusion of a planet operating in a mutually caring symbiotic relationship with humans; they remain anthropocentric and fail to undo the human as the logical cause for the earth's existence (Cohen, Colebrook and Hillis Miller 2012). To think interconnectedness without starting points of human or nature, of plurality or singularity, is a very difficult task indeed.

Security in a fragmented world

The inability of ecological security approaches to break free from dualism is telling. It points to the enduring power of a particular way of ordering the world in terms of problems and solutions, with those solutions framed in terms of escape to an outside (Walker 2010). In the ecological case, in explicitly claiming to move away from such binary thinking, its re-inscription is particularly problematic because it is obscured.

We must be wary of treating ecological security as offering an escape from the problems of modern politics. Wary, that is, of the temptation to respond to the demands of the Anthropocene by seeking a simple transcending or erasure of the boundary between the human and natural which might finally allow us to escape the violences entailed by such boundary drawing. The desire for a non-violent security politics cannot be fulfilled by attempting to transcend the violent system of oppositions and exclusions inscribed in the human/nature dualism, because to theorise an outside to this is to inscribe an oppositional system once again; to posit an interconnected whole as an alternative is itself a violent move.

Rather than seeking a more inclusive and differentiated account of security, on the basis of a biosphere with harmonious interests, the Anthropocene might be better engaged by a return to the contestations of politics and a foregrounding of the *negotiation* of representation, equality and domination, amongst an expanded constituency. One example of such an approach can be seen in Latour's (2004) argument for an expanded understanding of political representation in which the non-human can be included, and the need for rethinking political categories that this entails (Youatt 2014).

Without the assumption of ecological symbiosis and its underlying anthropocentrism, claims that security thinking can adapt to the demands of a world understood in terms of the Anthropocene become rather difficult to sustain; security understood in terms of protection or stasis no longer makes sense, but nor does security as resilience. As I have shown above, discourses of ecological security are in fact one mechanism by which dualist accounts of human and nature are reproduced. What is being secured here is precisely the human/nature binary, and the corresponding accounts of inclusion and exclusion, inside and outside, that this relies upon. The whole of a shareable world underpins all attempts to think outside, and it is this one, whole, shareable, world which is unsettled by the Anthropocene. In putting into question the human/nature distinction, the Anthropocene also puts into question both the possibility and desirability of security.

Rather than seeking a new – ecological – logic by which to domesticate the fractured world of the Anthropocene, embracing its reframing of our concepts of the 'world' might be a fruitful way with which to engage it. There can be no simple erasure or transcendence of the organising logic which has allowed us to conceive of ourselves as inhabiting an Anthropocene epoch. Rather than attempting to escape such a logic we might consider the Anthropocene instead to offer a framing of our political landscape which offers scope for a re-immersion in analysis of it. The Anthropocene might remind us that our assumption of a world as a condition of meaningful life and communication is, in Derrida's (2009) terms, a fiction, and so allow us to be attentive to the construction of that fiction and to the papering over the fissures and disjunctures within it required for its sustenance. By way of conclusion, I want to suggest two possibilities by which these breaks might be recognised, inhabited and opened up for political debate, through conceiving of them as cracks in the contextual field in which security and ecology operate.

Firstly, while the central contention of the chapter has been that employing the framework of security offers a very limited engagement with the challenges of the Anthropocene, there may still be a strategic, or pragmatic, case for pursuing the route of ecological security. The logics of security and the modern subject that I have identified as common to ecology and security are dominant, rather than necessary logics. Ecological thought offers compelling resources for theorising complexity, a lack of human mastery of the world and a critique of anthropocentrism.

However, as a logic, and especially when paired with security discourses, ecology/security also operate under a broader framework which acts to domesticate such complexity, to secure ecology as the logic of earth as homeland. Such a securing might well be considered a desirable – or at least a strategic, interim – aim, despite the fact that in and of itself it does not offer an alternative security logic. More broadly however, the tension *within* ecology between its content (complexity, lack of mastery) and form (logic) is precisely the type of fault line whose exploration might prove productive. If ecological security is successful in engaging security debates more broadly, then its dislocations and fractures may offer some starting points for changing the contextual field within which security logics need to make sense.

Secondly, questioning the possibility – but more importantly the desirability – of escape opens up space to try to think at the limits of the problematic oppositions identified here, rather than attempt to transcend them. The Anthropocene does not offer an escape from, or alternative to, the problems inherent in the human/nature dichotomy, it rather opens the possibility of thinking creatively within them; to return to Walker's terms, to think the way in which we are located always already within a relation between here *and* beyond, rather than to posit the beyond as escape. The Anthropocene shows us that we cannot escape the violences of oppositions and exclusions by a simple move to a shareable world. Rather than overcoming the human/nature binary, an Anthropocene sensibility highlights instead the way in which that boundary is multiplied, re-inscribed, magnified, produced, obscured and, most importantly, mobile (see Fagan 2019). It tasks us with inhabiting and re-examining its operation and the outsides that it (re)produces rather than trying to escape it.

With this in mind, while the failure of ecological security to transcend the anthropomorphism associated with the human/nature opposition is problematic on ecological security's own terms, it can be reframed as an invitation to reconsider anthropomorphism itself. Rather than offering an account of the changing context (or relationships) in which the human operates, there is scope for the Anthropocene to offer a vantage point from which to reimagine the very notion of 'human-in-relationship'. To return to Nancy's formulation, this might enable instead a consideration of the human *as* relation. As the etymology of the 'Anthropocene' suggests, the human might be new, open to change and negotiation, (re)produced by its 'new place'. Work on animal studies, posthumanism, cyborgs and new materialisms offer some routes that such a reimagining might take (see, for example, Bennett 2010; Braidotti 2013; Haraway 2015; Vaughan-Williams and Lundborg 2012; Wolfe 2012). This is in stark contrast to the modern subject of security, even that subject considered as in relation.

By putting into question the whole on which the human/nature divide relies the Anthropocene does not so much offer an alternative to it as loosen its anchor points, allow us to explore how its contours shift, and uncover the work that goes into stabilising them. We might consider, for example, the power practices that different iterations of the human/nature distinction make possible in the context of what Coole (2013) has identified as an extension of the commodification of nature to include elements *within* the human body – bacteria, genes, DNA – in such a way as to enable novel modes of intervention and management of that subject.

Refiguring the world in light of the Anthropocene makes it very difficult to relate to the planet, biosphere or ecosystem as a whole, but offers the opportunity to bring into focus instead the 'radical fragmented dissymmetry' between, and amongst, nature and humanity (Cohen, Colebrook and Hillis Miller 2012). To start from the position that we do not inhabit the same worlds as one another, that each living thing is finite, its world inaccessible to any other (see Derrida 2009; Nancy 2005) might then allow us to conceive of relation in a way which is not subsumed under, or determined by, a prior unity. There is certainly nothing secure about such relation; its negotiation is inherently political. The Anthropocene offers a potential opening to this risky venture; to rethink politics as something other than security politics, as a politics of vulnerability.

References

Barnett, J. (2001) *The Meaning of Environmental Security: Ecological Politics and Policy in the New Security Era*. London: Zed Books.

Bennett, J. (2010) *Vibrant Matter: A Political Ecology of Things*. London: Duke University Press.

Braidotti, R. (2013) *The Posthuman*. Cambridge: Polity.

Cohen, T., Colebrook, C. and Hillis Miller, J. (2012) *Theory and the Disappearing Future: On de Man, on Benjamin*. Abingdon: Routledge.

Colebrook, C. (2012) Not symbiosis, not now: why anthropogenic change is not really human. *Oxford Literary Review* 34(2): 185–209.

Coole, D. (2013) Agentic capacities and capacious historical materialism: thinking with new materialisms in the political sciences. *Millennium: Journal of International Studies* 41(3): 451–469. Corry, O. (2014) From defense to resilience: environmental security beyond neo-liberalism. *International Political Sociology* 8(3): 257–274.

Crutzen, P. J. (2002) Geology of mankind: the Anthropocene. *Nature* 415: 23.

Crutzen, P. and Stoermer, E. (2000) The Anthropocene. *Global Change Newsletter* 41: 17–18.

Cudworth, E. and Hobden, S. (2011) Beyond environmental security: complex systems, multiple inequalities and environmental risks. *Environmental Politics* 20(1): 42–59.

Dalby, S. (2009) *Security and Environmental Change*. Cambridge: Polity.

Derrida, J. (2009) *The Beast and the Sovereign: Volume I*. Chicago: University of Chicago Press.

Deudney, D. (1990) The case against linking environmental degradation and national security. *Millennium: Journal of International Studies* 19(3): 461–476.

Dobson, A. (2006) Do we need (to protect) nature? In J. Huysmans, A. Dobson and R. Prokhovnik, (eds), *The Politics of Protection: Sites of Insecurity and Political Agency*. Abingdon: Routledge.

Eckersley, R. (2007) Ecological intervention: prospects and limits. *Ethics & International Affairs* 21(3): 293–316.

Fagan, M. (2019) On the Dangers of an Anthropocene Epoch: Geological Time, Political Time, and Post-Human Politics. *Political Geography* 70: 55–63.

Floyd, R. (2010) *Security and the Environment: Securitisation Theory and US Environmental Security Policy*. Cambridge: Cambridge University Press.

Hamilton, C. (2015) Getting the Anthropocene so wrong. *The Anthropocene Review* 2(2): 102–107.

Haraway, D. (2015) Anthropocene, Capitalocene, Plantationocene, Chthulucene: making kin. *Environmental Humanities* 6: 159–165.

Homer-Dixon, T. F. (1991) On the threshold: environmental changes as causes of acute conflict. *International Security* 16(2): 76–116.

Kaplan, R. D. (1994) The coming anarchy: how scarcity, crime, overpopulation, tribalism, and disease are rapidly destroying the social fabric of our planet. *The Atlantic*, February. Accessed at: https://www.theatlantic.com/magazine/archive/1994/02/the-coming-anarchy/304670/.

Klare, M. T. (2001) *Resource Wars: The New Landscape of Global Conflict*. New York: Henry Holt.

Latour, B. (2004) *Politics of Nature: How to Bring the Sciences into Democracy*. Cambridge, MA: Harvard University Press.

Litfin, K. (1999) Constructing environmental security and ecological interdependence. *Global Governance* 5: 359–377.

McDonald, M. (2012) *Security, The Environment and Emancipation: Contestation over Environmental Change*. Abingdon: Routledge.

McDonald, M. (2013) Discourses of climate security. *Political Geography* 33: 42–51.

Mitchell, A. (2014) Only human? A worldly approach to security. *Security Dialogue* 45(1): 5–21.

Nancy, J. (1996) *Being Singular Plural*. Stanford: Stanford University Press.

Nancy, J. (2005) The insufficiency of 'values' and the necessity of 'sense'. *Journal for Cultural Research* 9(4): 437–441.

Pirages, D. (2013) 'Ecological security: a conceptual framework. In R. Floyd and R. Matthew (eds), *Environmental Security: Approaches and Issues*. Abingdon: Routledge.

Plumwood, V. (2002) *Environmental Culture: The Ecological Crisis of Reason*. London: Routledge.

Rudy, A. and White, D. (2013) Hybridity. In C. Death (ed.), *Critical Environmental Politics*. Abingdon: Routledge.

Trombetta, M. J. (2008) Environmental security and climate change: analysing the discourse. *Cambridge Review of International Affairs* 21(4): 585–602.

Vaughan-Williams, N. and Lundborg, T. (2015) New materialisms, discourse analysis, and international relations: a radical intertextual approach. *Review of International Studies* 41(1): 3–25.

Von Lucka, F., Wellmann, Z. and Dietz, T. (2014) What's at Stake in Securitising Climate Change? Towards a differentiated approach. *Geopolitics* 19(4): 857–884.

Walker, R. B. J. (1997) The subject of security. In K. Krause and M. Williams (eds), *Critical Security Studies: Concepts and Cases*. Abingdon: Routledge.

Walker, R. B. J. (2006) On the protection of nature and the nature of protection. In J. Huysmans, A. Dobson and R. Prokhovnik (eds), *The Politics of Protection: Sites of Insecurity and Political Agency*. Abingdon: Routledge.

Walker, R. B. J. (2010) *After the Globe, Before the World*. Abingdon: Routledge.

Wolfe, C. (2012) *Before the Law: Humans and Other Animals in a Biopolitical Frame*. Chicago: University of Chicago Press.

Youatt, R. (2014) Interspecies relations, international relations: rethinking anthropocentric politics. *Millennium: Journal of International Studies* 43(1): 207–223.

4

THE END OF RESILIENCE?

Rethinking adaptation in the Anthropocene

David Chandler

Introduction: the closure of time and space

Resilience has rapidly spread throughout the policy world over the last two decades, driven by the desire to use systems theories and process understandings to develop adaptive approaches for conserving, saving, sustaining and slowing down the runaway processes of modernisation and industrialisation. However, this chapter argues that under the auspices of the Anthropocene, the assumptions and goals of resilience become problematised. This is because the Anthropocene is held to close off the possibility of the spatial or temporal displacement of problems. In the Anthropocene we are told 'there is no away' and the consequences of what we do 'stick with us' (Morton 2013). In modernity, supporting and enabling vulnerable communities and ecosystems can help resolve problems but in the Anthropocene resilience approaches can easily appear to be spreading, rather than containing, the problem. Any attempt to resolve problems through focusing upon enabling and capacity-building can be seen to speed up the process of resource depletion and the arrival at the Earth's 'Planetary Boundaries' rather than slowing it down (Stockholm Resilience Centre n.d.). Even the more 'critical' and 'community-based' approaches, relying upon interventions to enable so-called 'natural' processes constitute further problems for resilience advocacy. Firstly, the problem of unrecognised exploitation and, secondly, the problem of continuing to sacrifice others to maintain the unmaintainable: Western modes of consumption and production.

Resilience approaches discursively frame policy problems and their resolution, through the focus on enabling and capacity-building communities and systems – held to be 'vulnerable', 'at risk' or 'failing' – through an imaginary that somehow natural, innate or inherent resources and productive capacities can be enhanced and developed. These potential imaginaries of resilience – as a policy-making 'magic bullet' for problems as diverse as underdevelopment, conflict and environmental crises – have come under challenge in the Anthropocene. This is because the Anthropocene is held to close off the spatial and temporal imaginaries of sustainable development and human progress. Resilience approaches require an outside of time and space to enable developmental progress to more sustainable solutions, both in the sense of the more efficient material production of necessities and in the assumption of reasoning capacity to learn and develop policy wisdom in a complex and interdependent world.

The argument of this chapter is that approaches attuned to the centrality of the Anthropocene provide a critique of resilience-thinking which is much more powerful than that levelled by critical societal and political theorists who have, over the last decade, condemned resilience discourses for their imbrications within neoliberal paradigms (Walker and Cooper 2011; Evans and Reid 2014; Chandler 2014; Joseph 2013). In fact, the Anthropocene appears to directly challenge the assumptions about resilience, neoliberalism and complexity developed by Jeremy Walker and Melinda Cooper (2011). Walker and Cooper argued that resilience-thinking was immune to critique, 'reabsorbing' or 'metabolizing critique into its internal dynamic' as 'the complex adaptive system remains self-referential even when it encounters the most violent of shocks' (2011: 157). The Anthropocene, with its promise of catastrophic global warming, climate change, species extinction and ocean acidification, would surely count as one of the 'most violent of shocks' imaginable. However, Walker and Cooper are not entirely mistaken in their assumptions about the impregnable nature of resilience-thinking. They conclude their piece with the words:

> In its tendency to metabolize all countervailing forces and inoculate itself against critique, 'resilience thinking' cannot be challenged from within the terms of complex systems theory but must be contested, if at all, on completely different terms, by a movement of thought that is truly counter-systemic.
>
> *(Walker and Cooper 2011: 157)*

For the purposes of this chapter, the Anthropocene is not constructed as merely one more 'shock' for the system, no matter the magnitude, but rather as precisely the grounds for a 'movement of thought that is truly counter-systemic'. As long as policy-makers and academic theorists presumed a modernist 'world' external to us and amenable to governing and policy interventions, resilience-thinking could 'reabsorb' or 'metabolize' shocks and 'bounce-back' through learning from disasters – even reimagining catastrophes as 'emancipatory' (Beck 2015) – or as facilitating new forms of self-growth and improved systems of self-management, 'bouncing-forward' with what former President of the Rockefeller Foundation, Judith Rodin, describes as the 'resilience dividend' (Rodin 2015).

However, the Anthropocene closes off the outside of time and space – a condition that is possibly easier to say or assert than to explain. In many ways, conceptualisations of the Anthropocene follow through on the promissory notes of the globalisation discussions of the 1990s in drawing out the implications of relational and system-thinking, which hold that there is no longer an 'inside' and an 'outside' (for example, Beck 1992). Constructions of 'insides' are those of autonomous agency or actors, central to modernist framings of law and politics, 'individuals', 'states', 'minds' etc are no longer conceived to be separate from the world of relations which constitute them, but as integral parts. 'Outsides' were seen to be merely mechanical, bound by natural laws and processes and amenable to objective knowledge and regulatory control, thus 'nature', the 'environment' and 'non-humans' were constructed as objects to be known by subjects. This division between insides and outsides enabled modernist imaginaries of 'progress', 'civilisation' and 'development', based on the intensification of these binary divisions. Fixed understandings of time and space were thus key to the constitution of the binaries of 'inside' and 'outside', forming the framework or backdrop, which was fixed and stable, within which human dramas were played out.

The Anthropocene enables 'a movement of thought that is truly counter-systemic' because time and space enter and thereby destabilise the idea of a separate 'inside'. As Amitav Ghosh powerfully notes, expectations of normality, balance and order that defined the modern world view appear from today's vantage point to be a terrible error or hubris: as carried to the point of 'great derangement' (Ghosh 2016: 36). There is a contemporary consensus that: 'There can be no more talk of linear and inexorable progress' (Bonneuil and Fressoz 2016: 21). As Bruno Latour argues:

> What makes the Anthropocene a clearly detectable golden spike way beyond the boundary of stratigraphy is that it is the most decisive philosophical, religious, anthropological and … political concept yet produced as an alternative to the very notions of 'Modern' and 'modernity'.
>
> *(Latour 2013: 77)*

The Anthropocene, initially a geological concept, claims that human actions have deeply affected and altered geologic processes, destabilising earlier 'Holocene' conditions of stability. Thus, we are threatened with catastrophic climate change not as some sort of 'external' threat to our modernist 'internal' narratives of sustainable development and human progress but precisely because our 'internal' understandings of humanity as somehow separate to, or above, the world, in a fixed and empty framework of time and space (with ourselves at the centre) were entirely false. Our stories of 'progress' and 'development' – the stories of our separation from nature – ignored the fact that we were actually destroying the very foundations of our planetary survival. This is why Anthropocene thinkers argue that the Anthropocene is not just another problem or crisis to be 'solved' or 'bounced-back' from or 'recouped' but rather a sign that modernity was a false promise of salvation, one that has brought us to the brink of destruction (Latour 2013; Stengers 2015; Tsing 2015).

So, while resilience-thinking has achieved nearly universal success in the policy-making world – suggesting new sensitivities to problems and rejecting 'high-modernist' technocratic approaches, which depended upon universal 'one-size-fits-all' solutions from on high – resilience is still a 'modern' construction which assumes that problems are 'external' and that we need to develop 'internal' policy solutions to maintain and to enable our existing modes of being in the face of shocks and perturbations. 'We' need to be more responsive and adaptable. 'We' need to be sensitive to minor changes and to 'tipping points'. In short, that 'we' are not the problem, but that 'we' need to develop new approaches to preserve our modernist imaginaries of development and progress. Resilience seeks to fight or to evade the conceptual assumptions of the Anthropocene rather than to accept them. The argument of this chapter is that precisely because the Anthropocene is not just an environmental threat but also the basis for a 'movement of thought that is truly counter-systemic', resilience-thinking has no weapons with which to successfully engage the Anthropocene ideationally. While resilience-thinking could 'reabsorb' or 'metabolize' critiques framed through modernist assumptions of overcoming and problem-solving, it is unable to 'adapt' to the new and increasingly prevalent subjectivities, sensitivities and imaginaries generated by the Anthropocene.

The claims staked out above, regarding the inability of resilience discourses to 'recuperate', 'reabsorb' or to 'metabolize' the criticisms of resilience approaches generated by Anthropocene sensitivities will be developed in the sections below. The next section explains the dominant critique of resilience-thinking today, that it seeks to 'defer' or evade underlying problems rather than to address them. With the closing off of modernist solutions, resilience appears to be 'coerced' or to be 'artificial', somehow lacking a genuine or real understanding of the limits to maintaining the status quo: this artificiality is sometimes expressed as problems of technical, bureaucratic, depoliticised or 'top-down' approaches which seek to achieve short-term solutions or to paper-over the cracks. Resilience solutions unintentionally cascade problems rather than ameliorating them. Then follows a section dealing with 'alternative' approaches to resilience, which seek to demonstrate the genuine or real capacities of resilience to grasp problems at source. These are framings of 'soft', 'natural', 'critical' or 'community' resilience. However, because the problem is constructed as 'external' to the policy interveners and as 'internal' to the community seen to require resilience, these alternative framings still fail to recuperate critique in the Anthropocene. In fact, the radicalisation of resilience approaches merely highlights the limits of modernist imaginaries that 'we' can still enable 'them' to 'bounce-back' when contemporary sensitivities suggest that it is precisely 'our' current modes of being and of perceiving our relationship to the environment which is the problem.

Coerced resilience: the problem of anthropogenic inputs

The problems that the Anthropocene posits for resilience advocacy have been little recognised in contemporary academic discussions in the humanities and social sciences. Resilience can appear to be still alive and well, if not actually thriving in policy debates centred around global warming and climate change. In fact, for the Stockholm Resilience Alliance – in the view of many commentators, the leading research and advisory body for resilience-thinking – the conceptualisations of resilience and of the Anthropocene are closely interconnected. Particularly in the language of systems ecology, both concepts appear to share understandings of complex adaptive systems, 'tipping points' and 'phase transitions' and to be sensitive to the limits of 'top-down' or 'linear' approaches to problem-solving. A glance at the Resilience Alliance webpages[1] reveals the clear interconnections between leading natural and social scientists, whose shared work in systems theory and adaptive systems has shaped thinking in both these areas, including Will Steffen, Paul Crutzen, Frank Biermann, Carl Folke, Johan Rockström and Jan Zalasiewicz, among others (see also Biermann et al. 2012; Steffen et al. 2011).

Yet, even at the 'heart of the beast' not all is well. One example of the limits of resilience-thinking comes from a group of Swedish ecology scientists linked with the Resilience Alliance (Stockholm Resilience Centre 2014) and published in *Ecosphere*, the journal of the Ecological Society of America (Rist et al. 2014). These scientists argue that resilience-thinking has been slow to think through the implications of the Anthropocene and the hidden costs of 'anthropogenic impacts on the environment'. The problem of ignoring these hidden costs is highlighted in their conceptualisation of 'coerced resilience', which they define as:

> Resilience that is created as a result of anthropogenic inputs such as labour, energy and technology, rather than supplied by the ecological system itself. In the context of production systems, coercion of resilience enables the maintenance of high levels of production.
>
> *(Rist et al. 2014: 3)*

Rist et al. define 'anthropogenic inputs' as the external 'replacement of specific ecosystem processes by inputs of labor and manufactured capital (e.g., fossil fuel, technology, nutrients, pesticides and antibiotics)' (Rist et al. 2014: 73). Thus sustaining or maintaining growth depends upon the taking of resources, technologies and materials from elsewhere, merely intensifying and redistributing or spreading the problems. This is, firstly, because the process is held to weaken and undermine 'natural processes' of resilience and, secondly, because importing resources weakens other, external, ecosystems.

Anthropogenic inputs make the problem worse by weakening rather than strengthening natural ecosystem sources of resilience. For Rist et al, this can be clearly seen in the shift to anthropogenic dependencies: with the development of intensive agriculture techniques over a thousand years ago; in forestry, which has moved to the industrial scale over the last few hundred years; and in fisheries, which became industrial after the Second World War (Rist et al. 2014: 4). In modernity, the problem was understood to be the ability to sustain these vulnerable systems, particularly with concerns of falling productivity. But in Anthropocene-thinking resilience itself becomes the enemy of resilience as the addition of anthropogenic inputs begins to shift the system regime state, moving further and further away from reliance on the natural ecological processes – and, in fact, causing permanent damage to them – until a new regime state is reached without the possibility of any return to 'nature' (Rist et al. 2014: 5). Thus vulnerabilities are cascaded through the larger system.

Rist et al. argue that one of the key problems with coerced resilience is that it 'masks' or hides the real costs of production through the import of external capital, namely in the form of technology and fossil fuel based energy (2014: 3). Thus the problem of some resilience policy interventions to enable sustainable development and human progress is thereby their 'artifice' or falsity. For some authors, this is akin to rearranging the deckchairs on a sinking ship as this merely takes materials from other ecosystems and contributes to spreading the problem rather than resolving it. In fact, coercive resilience is a kind of globalisation in reverse, where the ability to import goods from around the globe no longer adds to productivity but rather spreads the sickness of undermining natural processes by over extraction in unsustainable ways. For Rist et al. this 'falsity' is itself a key problem of coercive resilience, as it undermines the very feedback processes that complex adaptive systems require. In order to be productive, these systems:

> rely on the maintenance of local ecological processes to retain a wider range of options for unforeseen future requirements, and thereby provide clearer feedbacks regarding proximity to ecological thresholds than do production systems … which require significant anthropogenic inputs.
>
> *(Rist et al. 2014: 4)*

Thus increasing resilience through 'coercion' merely enables tipping points to be reached sooner. The addition of anthropogenic inputs 'masks' the growing loss of natural ecological system resilience maintaining systems in 'artificial' states, entirely dependent upon more and more external inputs:

> This raises an apparent paradox, whereby highly modified production systems can, through anthropogenic efforts rather than ecological processes, mimic the response of resilient natural systems to a specified disturbance, in their capacity to return to pre-disturbance system states.
>
> *(Rist et al. 2014: 6)*

This is a dangerous situation as artificial or 'coerced' resilience hides the capacities of these systems to draw upon natural ecological processes (highlighted in discussions of recent declines of wild and domestic pollinators and the plants and other species which rely upon them; see Rist et al. 2014: 6). A striking example of the limits of forced or coerced resilience is provided by anthropologist Michael Taussig (2018), in his recent work, *Palma Africana*, on the mass production of palm oil in Colombia. One of the unintended and ironic consequences of increasing reliance on anthropogenic inputs, for example, the development of mono-crops, such as the 'Hope of America' palm, is that although artificially designed to prevent the spread of insect predation it needs additional anthropogenic interventions to artificially inseminate it. Thus production becomes increasingly artificial, requiring more and more inputs, despite being sold as a wonderful technical solution for raising productivity:

> I see these women inseminators hard at it in the lustrous photographs provided by the Colombian Palm Growers Association. One woman is kneeling by an adult palm with a plastic tube in her mouth blowing sperm into the tiny flowers. In another photo a dark-skinned young woman wearing bright pink jeans and a coal black jacket and cap guides the inseminating tool in her right hand while with her left she pushes back the palm branches studded with fierce thorns. With a look of equally fierce concentration she guides her instrument into its target all because 'Hope of America' can't get it up. One would hope for more from 'Hope of America'.
>
> *(Taussig 2018: 74)*

In language which very much follows the lines of Rist et al., Taussig writes that:

> Once triggered, assemblages tend to proliferate and somersault, one leading to the next … Another assemblage concerns the larger framework of relevant political cliché and self-awareness as to such – namely, third world women of color ministering to the sexual requirements of an impotent masculine 'Hope of America' designed to stall the plagues brought by the very act of mono-cropping. We could continue. Thus does the assemblage principle provoke movement, speed, and metamorphosis. This is the way of things as much as a way of thinking with things.
>
> *(Taussig 2018: 75)*

Thus resilience, rather than halting or slowing down the process of environmental destruction and exhaustion, can in fact be seen as the very vector of its becoming. What is there to stop resilience from being retrospectively read into precisely the history of modernist developmentalism that it sets out to produce an alternative to?

For Rist et al., coerced resilience cascades system effects of resource depletion through increasing 'cross-boundary interactions' spreading the problem globally. One example they provide is that of livestock production, initially dependent upon farm-based resources and recycling waste products. In today's globalised interdependent world there is a decoupling of these processes, farm waste leaches into the environment rather than being recycled and intensive food production elsewhere (like soybean or palm oil) depends on ever higher inputs of synthetic mineral fertilisers, while global transportation merely adds to the consumption and waste of resources (Rist et al. 2014: 6–7). Thus vulnerabilities cascade through systems of positive-feedbacks, magnifying and extending the crisis of sustainability.

In the Anthropocene, it appears that any attempts to start from resilience 'problem-solving' assumptions can merely make the initial problem worse. This transformation or 'transvaluation' of solutions, which were previously seen as part and parcel of 'sustainable development' and 'progress', is due to the closing off of time and space. In the Anthropocene there is no 'outside' from which to draw resources. Modernity – now recast as the development of anthropogenic forms of 'cheating' nature – reaches its closure at a global scale, making coercive resilience not just the last gasp of modernity but actually the driver for its demise: 'because continued inputs are largely dependent upon, and ultimately limited by globally finite resources, such as fossil-fuel energy and phosphorous' (Rist et al. 2014: 7). The Anthropocene thereby spells the death knell for coerced resilience precisely through revealing the problem of 'masking' the environmental implications, which the distances of time and space had previously concealed. High levels of production and the speed of 'bounce-back' through resilience approaches were not enabling adaptation to new conditions but quite the opposite: merely working to 'mask or camouflage the ecological signals of resilience losses and thus the true underlying constraints to production' (Rist et al. 2014: 8).

Resilience is part of the problem not part of the solution. You don't have to be a scientist of system ecology (the original home of resilience-thinking) to realise that the days of resilience are thereby numbered. Criticisms of resilience for its artifice and lack of attention to the 'true underlying constraints' of modernity are 'cascading' across the academic disciplines. Resilience-thinking rather than being constructed as a challenge to modernist aspirations of 'command–and–control' is more likely to be seen as the last redoubt of eco-modernisers and of modernist dreams of technological and technocratic approaches which attempt to short-cut problems rather than to tackle them at source (for example, Schmidt 2013; Tierney 2015; Yusoff 2018; Neyrat 2019).

In the radical architecture journal *Places*, Lizzie Yarina provides a useful example, in her piece 'Your Sea Wall Won't Save You' (Yarina 2018). Drawing on research on the management of flood risks in Jakarta, Manila, Ho Chi Minh City and Bangkok, she argues that discourses of climate resilience seek to preserve the interests of global investors and manufacturing supply chains through enforcing technocratic and top-down solutions for unsustainable growth:

> Since the city [Bangkok, but it could just as well be Jakarta or elsewhere] occupies a flood plain, it will inevitably be inundated again; adding concrete just makes the problem worse. Bangkok is constantly fortifying against flood risk caused by its very existence, thus exemplifying Ulrich Beck's framing of a 'risk society' in which modern humans are caught in a cycle of mitigating risks they themselves have created.
>
> *(Yarina 2018)*

She advocates 'critical' or 'soft' approaches to resilience, based on the inclusion of communities rather than 'hard systems' of infrastructure and top-down technological or technocratic interventions, arguing that solutions lie in scaling up informal housing communities with local knowledge, rather than demolishing informal slums and relocating people elsewhere. Thus, not only do the technical resilience solutions to flood risk make the problems worse on their own terms, in displacing local people with local knowledge, they cascade the problem, through removing alternative or 'natural' processes of resilience:

> In decades past (and still today in some rural parts of the country) Thai people lived in 'amphibious communities': for example, in 'raft houses' which float upwards on stilts during floods, or in villages built on two levels where upper walkways and living quarters can be used during the rainy season. But those adaptive patterns are disappearing even as climate risk grows.
>
> *(Yarina 2018)*

Thus the problems of 'coerced' resilience become clear, and we can see a growing consensus that resilience is far from unproblematic as a set of governing interventions. For the critics of resilience considered above, resilience is problematised but there is hope for an alternative: resilience can be done better. There is a promise that resilience can be practised inclusively, 'naturally', in 'bottom-up' or 'non-anthropogenic' and 'artificial' ways. This promise will be put to the test in the following section which suggests that the implications of the Anthropocene need to followed through and that in doing so, even 'alternative' approaches to resilience can be understood as contributing to the problem rather than mitigating it.

The productivism of critical resilience

Discoveries about the power of nature to be able to be 'enabled' or 'scaled-up' in different ways through policy interventions would seem to be the inevitable 'last gasp' of resilience approaches. But what would non-coerced or non-anthropogenic approaches to resilience look like? Even, from the get-go, they would appear to be highly problematic. For resilience even to exist as a policy intervention – even to begin to 'attempt to use natural processes to enhance system resilience' (Rist et al. 2014: 8) – would appear to assume at least some anthropogenic actions. As Rist et al. advocate, often 'techno-fixes' may be required in the short-term as part of the process of using and manipulating 'natural processes' (Rist et al. 2014: 8):

> In such cases where coerced resilience is desired, the impacts on supporting and recipient system resilience must be considered. We argue that the ultimate goal is to retain or enhance the provision of global production system resilience through bolstering natural supporting processes rather than an increased reliance on anthropogenic inputs.
>
> *(Rist et al. 2014: 9)*

The game is rather given away here. The problems vitiating this approach are clear in the quote above. Firstly, there is a clearly instrumental approach to 'natural processes', which are to be harnessed to support the existing status quo, thus 'the ultimate goal' is to support 'global production system resilience'. This has come to the fore particularly in experiments in 'rewilding' and new forms of environmental conservation, seeking to enhance and expand 'ecosystem services', geo – and bio-engineering nature to be more efficient (see, for example, Lorimer 2015). As Anna Tsing notes, these resilience imaginaries are all part of an 'ecomodernist' fantasy of the 'good Anthropocene' (Tsing 2017: 16). Even if this could be achieved, 'natural processes' would be further modified by anthropogenic manipulation: the mere need to intervene to 'bolster' these allegedly 'natural processes' would inevitably produce other unintended stresses and strains according to the logic of the authors' own arguments. As Frédéric Neyrat argues: 'saving the planet can only mean one thing, and this is one of the leitmotifs of post-environmentalism: *Intervene even more* – in other words. "creating and re-creating [the Earth] again and again for as long as humans inhabit it"' (Neyrat 2019: 85, italics in original). In the Anthropocene, arguments for the instrumental use of 'nature' can only speed up the process of catastrophic collapse. Secondly, there is the problem of the 'ultimate goal', which is the enhancement 'of global production system resilience'. Even if nature could be enrolled and manipulated through resilience policy interventions without the side effects of undermining these processes further, global production resilience is entirely the wrong goal in the Anthropocene. The system of global production itself is the problem not the solution!

The end of the resilience project is starkly clear in the examination of the imaginaries of 'soft', 'natural', 'non-coerced', 'community' or 'critical' approaches to resilience. Ironically perhaps, these last gasps of resilience-thinking produce some of the most backward, romantic and essentialising discourses of governance, perhaps fitting for the collapse of the aspirations of modernity, but often distasteful to read, especially when the promoters of these ideas consider themselves to be 'radical' or 'critical' in some way. Perhaps the worst aspect of 'critical' approaches to resilience in the time of the Anthropocene is that the costs and burdens of sustaining the world in the imaginary of the end of modernity are inevitably borne by those least able to resist the requirements of power.

The 'front-line' resisters of the impacts of climate change are always those communities constructed as marginal, as somehow closer to the problematic effects of environmental change, whether through subsistence modes of living or through other forms of economic and social vulnerability. Resilience discourses never fail to highlight the impact of climate change on precarious modes of being, yet, as Kyle Whyte (2019) notes, addressing vulnerability to climate change without reference to the larger context of capitalism and colonialism puts the emphasis for adaptive change on the victims while maintaining the systems responsible for the unequal distribution of effects. Resilience cannot possibly be undertaken in 'soft', 'critical' or 'community' ways without very clearly redistributing the burdens of risk and sacrifice. Even more problematically, these approaches inevitably assume that there is a hidden or cost-free resource that can be used, whether this is understood to be found in 'nature', in 'informal', 'indigenous' or 'non-modern' modes of being or in self-monitoring and self-policing communities coping on the edge of poverty. These 'alternative' approaches merely reproduce the problems of more 'technical' or 'top-down' coerced approaches in, firstly, ignoring the unintended or future costs and, secondly, and most importantly, assuming that modernist modes of consumption and production can continue practically unchanged.

Heuristically, the space in which 'critical' resilience approaches work can be clarified in terms of the need to operate through policy interventions which are 'non-anthropogenic' in intent, i.e. which seek to intervene in social, economic and ecological processes with the goal of enabling or drawing out 'natural' processes to enhance productivity and efficiency through tightening relational sensitivities and enabling better ways of sensing and responding to 'natural' 'signals' and 'feedbacks'. Thus resilience, as a set of policy interventions to enable adaptive capacities, always necessitates an instrumental or goal-directed set of sensitivities: natural or immanent processes cannot be 'enabled' by being left alone. Nature cannot ever be considered as somehow operating separately to social, economic and political processes. The starting assumption for resilience discourses is that we are now 'after Nature' (Purdy 2015; Lorimer 2015) or 'after ecology' (Morton 2009, 2013; Latour 2004). As Gleb Raygorodetsky argues, on behalf of the struggle to develop climate change adaptation and mitigation responses, even if we set aside half the planet as nature, as the Harvard biologist E. O. Wilson (2016) famously suggests:

> This strict stance, however, does little to help get to the root of our destructive behaviour. Allowing development to destroy habitat in one area with a promise of 'offsetting' this destruction by conserving another place actually perpetuates humankind's assault on the environment. It creates an illusion that as long as a portion of nature is put away and locked up in some sort of a park, we can rape and pillage the rest of the planet.
>
> *(Raygorodetsky 2017: 180)*

As the Brazilian architectural theorist Paulo Tavares argues, the Western idea of a pristine 'nature' that can be preserved or kept away from human interaction has always been mythical (Tavares 2013: 234). Even the Amazonian rain forests have been cultivated in sustainable ways by indigenous communities, thus 'Amazonia's deep history is not natural, but human' (2013: 234): 'And this is perhaps the crucial paradox that the Anthropocene has brought to light: different regimes of power will produce different natures, for nature is not natural; it is the product of cultivation, and more frequently, of conflict.' (2013: 236)

For critical approaches to resilience the alternative to mono-crop agriculture, industrialised fisheries, sea walls and river 'normalisation' is never to 'just let nature take its course'. In discursive framings that are little different to neoliberal constructions of governance interventions that are 'for the market' – designed to enable or to 'free' the productive and organisational capacities of market forces – 'nature' (like market forces) is never assumed to be 'natural' (see Chandler 2014). Nature, no longer separate to human systems, requires wise and active stewardship, like any other complex adaptive system. 'Critical' approaches to resilience are thereby not against technological applications and understandings but seek to apply them differently: to work 'with' rather than 'against' immanent productive processes, sensitive to feedbacks and unintended effects.

A review of the literature on 'critical', 'soft' and 'community' approaches to resilience suggests that there are two dominant approaches in the problem of developing sensitivity to feedbacks and thus avoiding the dangers of 'coerced' resilience. The first focuses upon the importance of local community knowledge and experience. In discourses of resilience two figures of community emerge which embody the types of adaptive knowledge required, both of which are constructed as non-modern or non-universalist ways of knowing and of becoming sensitised to feedback effects, gained through lengthy experiences of coping with contexts of difficulty and trauma. These are the figures of the informally settled community, attuned to environmental turbulence, and of indigenous communities, respectful of their relations to non-human others. Both these figures are imagined as resilient and self-sustaining communities, capable of coping, adapting to and 'bouncing back' from regular disturbances and disruptions.

There are a number of problems with these iconic figures of 'critical' and 'community' resilience. Although often well intentioned, it is difficult for Western agencies and activists to escape accusations that they are essentialising and romanticising the life-styles and coping strategies of the marginalised communities they are offering up as role models for adaptive approaches. Since the early 1990s, there has been a major policy shift from earlier slum removal to slum improvement, slum rehabilitation and slum development programmes (Davis 2006). Slum dwellings are increasingly high on international policy agendas, enabling the merging of security concerns of poverty, climate change and urban growth to be renegotiated through the lens of resilience. In these developing approaches, resilience is something that can be generated through engagement with urban slums and often through the application of new technologies for community engagement and local leadership (Castroni 2009; Ogunlesi 2016). If new forms of seeing relationally and responding to feedback effects are the model for policy intervention, then slum dwellers and the urban poor are the most proficient at organically developing solutions based on sensitivity to feedback. As the *Economist* notes:

> In a way, slums are areas of high sustainability – they use less water and electricity, for example. There is also a stronger sense of community and solidarity than in big cities in general, which are much more anonymous. Slum dwellers are particularly entrepreneurial, with families converting their ground floor into a soup kitchen or a school. Policymakers in developed cities should learn to listen to citizens rather than adopt a top-down approach to planning – a core component of the 'slum upgrading' method.
>
> *(Brillembourg 2015)*

Here, the approach is very different to the liberal grand schemes of social engineering and slum clearances or the neoliberal assumption that slum dwellers are, in some way, lacking capacities and in need of external agencies to provide them with resilience. While there are many good arguments against the forced resettlement of informal communities, often to areas without suitable community support structures, the idea that slums should be scaled-up and enlarged so that informal ways of coping can be put at the disposal of the city dwellers, enabling them to continue undisturbed, seems to be exploitative rather than emancipatory.

A similar set of arguments would appear to undermine some of the claims made by environmental campaigners, who insist that indigenous communities be enrolled as 'traditional custodians' in support of the latest Intergovernmental Panel on Climate Change (IPCC) reports (for example, Forest Peoples Programme 2018). The essentialist and exploitative framing which seeks to 'support' indigenous communities in maintaining biodiversity on 'our' behalf often leads to oppressive forms of regulation and categorisation where indigenous rights become dependent on community members pledging themselves to maintain their ancestral beliefs and practices (Andersen 2014; Sissons 2005; Simpson 2014; Tallbear 2013). As Sissons (2005: 37) suggests, 'indigenous authenticity is racism and primitivism in disguise'. These views are merely a Western revaluation of 'primitivism and tribalism in relation to destructive western rationality and individualism'. Rather than a discourse of indigeneity, Sissons suggests these views be understood as 'eco-ethnicity', where ecological threats are 'ethnicized' and ethnic subordination 'ecologized' (Sissons 2005: 23).

Apart from being romanticising and essentialising, a lot of the claims made on behalf of these marginalised communities do not stand up to close examination. As Nigel Clark (2011) notes, natural forces have always been dangerous, unpredictable and sometimes catastrophic, making views of attunement to natural rhythms and signs impossible regardless of harmonious imaginaries of indigenous forms of adaption and sustainability (205–6): 'all environmental knowledge is hard won, provisional and fallible' (207). In the epoch of the Anthropocene, local and indigenous knowledge is even less of a guarantee of effective responsiveness. In many ways it is ironic that although the interlocutors from informal and indigenous communities, that Western advocates draw upon, repeatedly state that they can no longer adapt in traditional ways – to changes in the river's path and momentum (Yarina 2018; Chandler 2017: 121) or that the climactic and seasonal signs that used to provide a guide to everyday life are now much more erratic and unreliable (Raygorodestsky 2017: 59) – the 'voices' of the people themselves are rarely heard in the rush to instrumentalise these survival strategies as 'critical' alternatives. It seems clear that what is being drawn from these communities says much more about the desires of Western advocates and activists than about these communities themselves, many of which are adapting to change (including the impacts of climate change) in ways which have increasingly less and less relation to traditional or local knowledge-based practices (Raygorodestsky 2017: 52, 193, 243).

Beyond the backward imaginaries of indigenous communities as 'testing grounds' and 'laboratories' (Raygorodestsky 2017: 258) there is often a slightly more sophisticated international policy agenda of understanding and scaling up adaptive capacities (UNESCO 2017). The capacity that local subsistence and indigenous communities are imagined to have (and Western societies are imagined to lack) is the ability to see and respond to feedback effects (for example, Lansing 2006). As botanist Robin Wall Kimmerer argues, in her work *Braiding Sweetgrass*, 'becoming indigenous' means 'to take care of the land as if our lives, both material and spiritual, depended upon it' (Kimmerer 2013: 9). Her emphasis is very much on resilience as an ethos of care and reciprocal becoming with others. The enemy of resilience is the breakdown of feedback loops in the artifice of modern society which tears away any possibility of reciprocity, creating a 'Potemkin village of an ecosystem where we perpetrate the illusion that the things we consume have just fallen off the back of Santa's sleigh, not been ripped from the earth' (Kimmerer 2013: 199).

As 'critical' approaches to resilience develop, in their ability to articulate resilience as an approach, they increasingly move beyond an understanding that local and traditional knowledge itself is key and towards the development of resilience as a methodology or as a set of feedback analytics (for example, Kimmerer 2013; Povinelli 2016) which can be exported and scaled-up on the basis of enabling communities to become resilient through being able to monitor and respond to changes in climactic and other conditions with greater speed and efficiency than waiting for government policy interventions. Thus the second approach to 'critical' or 'community' resilience focuses less upon autonomous local knowledge capacities and more upon how new technological advances in algorithmic computation and distributive sensory capacities can enable local communities to be more self-sustaining.

The use of technology, not as a 'techno-fix' that artificially hides feedback effects but rather as one that enables them to be seen and responded to, is now central to 'critical' resilience imaginaries in the Anthropocene. The rolling out of Big Data and the Internet of Things approaches to local communities promises a level of responsiveness and sensitivity to environmental changes that was previously unimaginable. For its boosters in international development agencies and corporations, these approaches will transform small-scale agricultural production. Even palm oil production receives a critical makeover. Rather than environmentally destructive industrial mono-cropping, small plot alternatives can be made economically viable if farmers sign up to digitally enhanced 'cloud-based' management systems, where farmers enable large scale data collection and sensory monitoring systems to be installed and so can monitor and minimise the use of chemicals and other anthropogenic resources as well as rapidly respond to drought, pests and disease – detecting problems even down to the level of specific trees and plots. Just as with Google and Amazon, sensitivities to feedbacks increases the more data is shared and drawn upon. As the founder of one agri-tech start-up states:

> We specifically use … cloud storage (to store raw and processed imagery), cloud compute (to process huge amounts of data and extract insights), database storage and to serve our applications … to help farmers grow healthier crops is a perfect example of the way in which technology transforms traditional industries, leading to better livelihood conditions. Africa can be a harsh

environment for farming. Crops are constantly under threat from problems such as disease, pests, and drought. Using the ... cloud, we are bringing computation, data analytics, and other advanced technologies to help farmers grow healthier crops, despite the harsh conditions.

(Cline 2018)

Just as 'coerced' resilience spread the problems of resource depletion and cascaded the lack of sustainability throughout the global system, it appears that 'critical' resilience spreads 'zones of sacrifice' scaling them up as one part of the world is called upon to be resilient while the other part of the world 'can rape and pillage the rest of the planet'. While earlier ecological approaches advocated that parts of the world should literally be preserved to enable the rest of the world to go on as before, 'critical' resilience approaches argue that local communities should be maintained in informal or indigenous modes of life to act as 'testing grounds' or 'laboratories' – as front-line 'responders' – in the Anthropocene. These separations and distinctions are now beginning to cascade up through 'critical' imaginaries that seek to monitor, police and regulate the most marginal communities that are told that they need to scale-up their capacities for resilience to enable others to continue producing and consuming as before.

Conclusion

Resilience is all about the capacity for adaptation to problems rather than traditional problem-solving. What is interesting about contemporary critiques of resilience, grounded in Anthropocene sensitivities, is that they assert that adaptation itself is problematic; precisely because adaptive approaches seek to sustain the system or modes of being that appear to be under threat from shocks or crises. For Anthropocene approaches, these systems themselves are the problem: enabling them to adapt, to sustain themselves or to thrive merely makes the problems worse. In short, the two forms of resilience considered here – both the 'technocratic' or 'coerced' and the alternative 'critical' or 'community' approaches – will inevitably be understood as counterproductive; seeking to evade the real nature of the problems posed by the Anthropocene and to redistribute costs and burdens increasingly upon those least able to resist these modernist imaginaries. Discourses of adaptation merely highlight the existing inequalities, exclusions and blind spots in modernist resilience-thinking. Rather than being seen as challenging the hubris of 'high-modernism', the Anthropocene casts discourses of resilience and adaptation as the last desperate attempts to hang on to these increasingly discredited practices and ideas.

Note

1   https://www.resalliance.org

References

Andersen, C. (2014) *Métis: Race, Recognition, and the Struggle for Indigenous Peoplehood*. Vancouver: University of British Columbia Press.

Beck, U. (1992) *Risk Society: Towards a New Modernity*. London: Sage.

Beck, U. (2015) Emancipatory catastrophism: what does it mean to climate change and risk society? *Current Sociology* 63(1): 75–88.

Biermann, F., Abbott, K., Andresen, S., Bäckstrand, K., Bernstein, S., Betsill, M. M., … Zondervan, R. (2012) Navigating the Anthropocene: improving earth system governance. *Science* 335(6074): 1306–1307.

Bonneuil, C. and Fressoz, J.-B. (2016) *The Shock of the Anthropocene*. London: Verso.

Brillembourg, A. (2015) Learning from slums. *Economist Intelligence Unit*, 28 January. Accessed at: https://www.eiuperspectives.economist.com/infrastructure-cities/learning-slums.

Castroni, M. (2009) Learning from the slums: literature and urban renewal. *Arch Daily*, 8 March. Accessed at: http://www.archdaily.com/15271/learning-from-the-slums-12literature-and-urban-renewal/.

Chandler, D. (2014) *Resilience: The Governance of Complexity*. Abingdon: Routledge.

Chandler, D. (2017) Securing the Anthropocene? International policy experiments in digital hacktivism: a case study of Jakarta. *Security Dialogue* 48(2): 113–130.

Clark, N. (2011) *Inhuman Nature: Sociable Life on a Dynamic Planet*. London: Sage.

Cline, T. (2018) Big data and smart farmers For Africa's agricultural transformation. *Forbes Africa*, 30 October. Accessed at: https://www.forbesafrica.com/focus/2018/10/30/big-data-and-smart-farmers-for-africas-agricultural-transformation/.

Davis, M. (2006) *Planet of Slums*. London: Verso.

Evans, B. and Reid, J. (2014) *Resilient Life: The Art of Living Dangerously*. Cambridge, UK: Polity Press.

Forest Peoples Programme (2018) November newsletter. *Forest Peoples Programme*, 1 November. Email communication.

Ghosh, A. (2016) *The Great Derangement: Climate Change and the Unthinkable*. Chicago, IL: University of Chicago Press.

Joseph, J. (2013) Resilience as embedded neoliberalism: a governmentality approach. *Resilience: International Policies, Practices and Discourses* 1(1): 38–52.

Kimmerer, R. W. (2013) *Braiding Sweetgrass: Indigenous Wisdom, Scientific Knowledge and the Teachings of Plants*. Minneapolis: Milkweed Editions.

Lansing, J. S. (2006) *Perfect Order: Recognizing Complexity in Bali*. Princeton: Princeton University Press.

Latour, B. (2004) *Politics of Nature: How to bring the Sciences into Democracy*. Cambridge, MA: Harvard University Press.

Latour, B. (2013) *Facing Gaia, Six Lectures on the Political Theology of Nature: Being the Gifford Lectures on Natural Religion, Edinburgh, 18th–28th of February 2013* (draft version 1–3-13). Accessed at: https://macaulay.cuny.edu/eportfolios/wa kefield15/files/2015/01/LATOUR-GIFFORD-SIX-LECTURES_1.pdf

Lorimer, J. (2015) *Wildlife in the Anthropocene: Conservation after Nature*. Minneapolis: University of Minnesota Press.

Morton, T. (2009) *Ecology without Nature: Rethinking Environmental Aesthetics*. Cambridge, MA: Harvard University Press.

Morton, T. (2013) *Hyperobjects: Philosophy and Ecology after the End of the World*. Minneapolis: University of Minnesota Press.

Neyrat, F. (2019) *The Unconstructable Earth: An Ecology of Separation*. New York: Fordham University Press.

Ogunlesi, T. (2016) Inside Makoko: danger and ingenuity in the world's biggest floating slum. *Guardian*, 23 February. Accessed at: http://www.theguardian.com/cities/2016/feb/23/makoko-lagos-danger-ingenuity-floating-slum

Povinelli, E. (2016) *Geontologies: A Requiem to Late Liberalism*. Durham: Duke University Press.

Purdy, J. (2015) *After Nature: A Politics for the Anthropocene*. Cambridge, MA: Harvard University Press.

Raygorodestsky, G. (2017) *The Archipelago of Hope: Wisdom and Resilience form the Edge of Climate Change*. New York: Pegasus Books.

Rist, L., Felton, A., Nystrom, M., Troell, M., Sponseller, R. A., Bengtsson, J., … Moen, J. (2014) Applying resilience thinking to production ecosystems. *Ecosphere*, 5(6), article 73.

Rodin, J. (2015) *The Resilience Dividend: Managing Disruption, Avoiding Disaster, and Growing Stronger in an Unpredictable World*. London: Profile Books.

Schmidt, J. (2013) The empirical falsity of the human subject: new materialism, climate change and the shared critique of artifice. *Resilience: International Policies, Practices and Discourses* 1(3): 174–192.

Simpson, A. (2014) *Mohawk Interruptus: Political Life across the Borders of Settler States*. Durham: Duke University Press.

Sissons, J. (2005) *First Peoples: Indigenous Cultures and their Futures*. London: Reaktion Books.

Steffen, W., Persson, Å., Deutsch, L., Zalasiewicz, J., Williams, M., Richardson, K., … Svedin, U. (2011) The Anthropocene: from global change to planetary stewardship. *Ambio* 40(7): 739–761.

Stengers, I. (2015) *In Catastrophic Times: Resisting the Coming Barbarism*. Paris: Open Humanities Press.

Stockholm Resilience Centre (n.d.) The nine planetary boundaries. *Stockholm Resilience Centre*. Accessed at: https://www.stockholmresilience.org/research/planetary-boundaries/planetary-boundaries/about-the-research/the-nine-planetary-boundaries.html

Stockholm Resilience Centre (2014) The hidden cost of coerced resilience: Centre researchers look into forced resilience of intensive agriculture, forestry, fisheries and aquaculture systems. *Stockholm Resilience Centre*, 29 November. Accessed at: https://www.stockholmresilience.org/research/research-news/2014-11-29-the-hidden-cost-of-coerced-resilience.html

Tallbear, K. (2013) *Native American DNA: Tribal Belonging and the False Promise of Genetic Science*. Minneapolis: University of Minnesota Press.

Taussig, M. (2018) *Palma Africana*. Chicago: University of Chicago Press.

Tavares, P. (2013) The geological imperative: on the political ecology of the Amazonia's deep history. In E. Turpin (ed.) *Architecture in the Anthropocene: Encounters among Design, Deep Time, Science and Philosophy*. Ann Arbor: Open Humanities Press, 207–239.

Tierney, K. (2015) Resilience and the Neoliberal project: discourses, critiques, practices – and Katrina. *American Behavioural Scientist* 59(10): 1327–1342.

Tsing, A. L. (2015) *The Mushroom at the End of the World: On the Possibility of Life in Capitalist Ruins*. Princeton, NJ: Princeton University Press.

Tsing, A. L. (2017) The Buck, the bull, and the dream of the stag: some unexpected weeds of the Anthropocene. *Suomen Anthropologi* 42(1): 3–21.

UNESCO (2017) *Local Knowledge, Global Goals*. Paris: UNESCO. Accessed at: http://www.unesco.org/new/fileadmin/MULTIMEDIA/HQ/SC/pdf/ILK_ex_publication_E.pdf

Walker, J. and Cooper, M. (2011) Genealogies of resilience: from systems ecology to the political economy of crisis adaptation. *Security Dialogue* 42(2): 143–160. Whyte, K. (2019) Way beyond the lifeboat: an indigenous allegory of climate justice. In K.-K. Bhavnani, J.

Foran, P. A. Kurian and D. Munshi (eds), *Climate Futures: Reimagining Global Climate Justice*. London: Zed Books.

Wilson, E. O. (2016) *Half-Earth: Our Planet's Fight for Life*. New York: Liverlight Publishing Corporation.

Yarina, L. (2018) Your sea wall won't save you. *Places Journal*, March. Accessed at: https://placesjournal.org/your-sea-wall-wont-save-you/

Yusoff, K. (2018) *A Billion Black Anthropocenes or None*. Minneapolis: University of Minnesota Press.

5

COLLIDING TIMES

Urgency, resilience and the politics of living with volcanic gas emissions in the Anthropocene

Sébastien Nobert, Harold Bellanger Rodríguez and Xochilt Hernández Leiva

Introduction

The recent images of Hawaii's Kelauea and Guatemala's Fuego volcanic eruptions have reiterated to the world that volcanoes are unpredictable forces of destruction. Their lava flows and pyroclastic clouds are indiscriminately reducing human, plant and animal lives to ashes, and are a strong reminder that their capacity to shape and reshape entire landscapes and life forms can occur at any time. While environmental hazards such as volcanoes have the capacities to reveal the vulnerabilities and failures of the project supporting the modern ideal, they are also undeniable evidence that what holds together our ontological security is challenged by the indeterminacies brought about by the unthinkable (Dupuy 2002). This crude reality is also helping us to rethink the Anthropocene era, as the planet has never been suborned to human needs and interests and it will never be (see Latour 2014, 2015; Stengers 2009, 2015 on this theme). Even the most sophisticated tools, such as early warning systems, policy instruments, such as the precautionary principle, or post-disaster strategies, such as resilience, are incapable of preventing and stopping the geopowers (e.g. geological processes) of our planet such as magmatic activity (see Grosz 2004, Yusoff et al. 2012).

By their nature and effects, geological processes – or geopowers – are dragging us outside the modalities of panic and urgency posed by hypermodernity and the eschatological threats of the Anthropocene. They connect us to the slow movements and processes that have been actualising continually the shape of the Earth since the metamorphosis of the Gondwana, passing through the little Ice Age, to contemporary earthquakes and soil movements that make the headlines of spectacle news. Volcanoes are signs of this geologic life that surrounds us and that we tend to ignore (Yusoff 2013), a life that connects us to the more fundamental questions related to the essence of time, both in the shape of *deep time* and of what we can define as the time of ephemeral human lives. Deep time refers to the inter-relationships that give shape to geological movement and difference and that are part of what Henri Bergson (1907) calls the *élan vital*, this vital force that defines the ontological dimensions of differences and that precedes and succeeds human life. Thus volcanoes are not only a source of spectacle and destruction, but they also inform us about the wider presence of deep time corporeality and in doing so, they expose its collision with the time produced and experienced by humans. This temporal collision is very central to the Anthropocene, as not only is its agency made of deep geological time, but also its trajectory is projecting us into the accelerated dynamics of anticipation and anxiety that define the modality of the future it purports.

This is perhaps the paradox about the Anthropocene as defined and engaged with by most social scientists in recent years (e.g. Bonneuil and Fressoz 2012; Hamilton et al. 2015) as it is posed as a problem requiring fast and immediate action while it is partly defined by deep geological time. It is the paradox of finding an urgent response to the Anthropocene, which demands speeding up what consists of the now in the hope of recasting our relationships with the non-humans in the future, but in ways that are tacitly carrying the expectation of being capable of synchronising social action through the single experience of accelerated time (see Rosa 2013). This results in it being difficult to situate political possibilities outside of the temporalities produced by and reflected through the accelerated futures of the modern ideal.

The geological tale that has worked to shape the Anthropocene corresponds to what the French phenomenologist Henri Maldiney (2012) calls *explicated time*, a time divided into the categories of the past, the present and the future; time as represented into periods and epochs. Thereby, explicated time offers the means to stabilise this never-ending-becoming that the concept of time attempts to capture; it provides a frame for the multiple trajectories embodying the movement of life. Blurring the contours of explicated time, Maldiney (2012) defines what he sees as *implicated time*, which corresponds to the time of experience, irreducible to metrology and explicated time categories. It is the time we feel, that occurs simultaneously with explicated time but that is so vital to our experiences and our daily lives that we have stopped questioning its emergence and substance. Implicated time is not dominated by the category of the future, but is rather made of multiple rhythms (i.e. the different speeds at which time is produced and experienced) and durations (i.e. contraction or extension of intersubjective time; the phenomenological expression of time that makes some events survive in the flow of the moment) that are leading to disruptions in time continuum (see Nobert et al. 2017; Nobert and Pelling 2017). By looking at the ways in which different sets of knowledge and practice are involved in shaping ways to deal with volcanic air pollution from Nicaragua's Masaya volcano, this chapter aims to document how the implicated temporal processes of rhythms and durations come to challenge the temporalities of resilience disaster risk reduction strategies (DRR) put forward. This empirical exploration makes it possible to highlight how temporal collisions and negotiations take shape between, on the one hand, everyday experiences of volcanic life within local communities, and, on the other hand, modern visions of resilient futures found within the wider DRR discourse.

Empirically this chapter draws on a set of 15 semi-structured interviews, three focus groups and field notes from observations undertaken with communities impacted by permanent volcanic emissions (PVEs) of Masaya volcano during the period covering February to April 2017. Borrowing on the social studies of disasters and risk literature, PVEs are not considered here merely in terms of their chemical composition and physical appearance as floating and hazy blue toxic gases; rather, they are conceptualised as geological elements through which temporal collisions and formations can be captured while providing a better understanding of the micro-politics involved in dealing with hazards that are too often ignored by DRR proponents.

After discussing our methodology and fieldwork in more detail, this chapter looks at the temporalities of resilience and stresses how time has been engaged with by disaster studies. This trajectory makes it possible for the chapter to show how the modern DRR quest for synchronicity in the production of resilient and medicalised futures is in turn demonstrating how different ways of experiencing time and movement have been ignored in the making and understanding of resilience. We show that this temporal exclusion is challenged by different ways of producing and experimenting rhythms and durations by those living with PVEs, highlighting how dealing with PVEs is inspired by the unexpected of the *now* rather than in the preformatted future of DRR resilience that consists in bouncing back to the state defined as normality. The chapter concludes by stressing that if the Anthropocene is to become an opportunity to rethink the place of humans in the realm of geological life, then what is conceived as time needs to be opened up to a plurality of ways of knowing and sensing non-human life that are implicated in the processes of dealing with hazards.

Doing field work in a paranoiac state

Doing fieldwork in Nicaragua is not an easy process.[1] As these lines are written, Nicaragua continues to be engulfed in political turmoil (see Phillips 2019), with a youth movement that aspires to freedom and fair elections and an oppressive political regime led by the Sandinista government and its presidential couple, Daniel Ortega and Rosario Murillo. The presidential couple can be found in all aspects of Nicaraguan life, ranging from Mrs Murillo's involvement in communicating TV morning news, to massive billboards showing the couple waving at the people and celebrating their 2016 victory, or the sight of the Sandinista flag floating above governmental buildings. More recently, though, the couple made international headlines for their handling of the April 2018 student uprising, which resulted in ordered assassinations of journalists and dissidents, deportations of foreign journalists, the disappearances of political activists, and the widespread incarceration of many young and old Nicaraguans (see Lakhani 2018). Such violence has not only resulted in a breach of human rights, but has also led to the migration of many nationals who, like the students, were opposing the return of a dictatorial regime to their country after decades of civil war.

Although our fieldwork predated the 2018 student uprising, the political grip of Ortega and Murillo was palpable throughout the research process and has impacted on the possibility of talking to people freely as well as on the ability to conduct a snowballing sampling method without interference from political officials. The omnipresent shadow of Sandinista in Nicaragua meant that performing semi-structured interviews and conducting focus groups needed to acknowledge this controlling reality. This peculiar situation explains why our number of recorded semi-structured interviews is limited. Those who agreed to be interviewed needed also to feel comfortable with being recorded and while research ethics forms were distributed and guaranteed anonymity to our interviewees, the presence of Sandinista informants in communities meant that several interviewees accepted to answer our questions but refused to be recorded. This new level of scrutiny meant that note-taking, along with visual methods such as photographing in addition to small talks, became important sources of triangulation conducted by the team of social scientists in order to validate the interviews' content. All interviews, focus groups and notes were transcribed into Spanish and discourse analysis was performed to find elements related to the research themes.

Uncovering the time of resilience in the social studies of hazards and risk

Keeping the focus on the not-yet

In the wider realm of DRR, both the materiality and temporality of PVEs pose important questions about the types of measure and strategy that should be prioritised to deal with them. This is because even when volcanoes are not erupting, their emissions can be extremely rich in acids (e.g. sulphur dioxide gas), fine particulate matter (e.g. PM2.5) and heavy metals (e.g. lead, arsenic) that have a great potential to affect air and water qualities and impact directly on the health and well-being of local communities (Baxter et al. 1982, 1983; Demelle et al. 2002; Van Manen 2014). The persistent nature of PVEs carries us into the geological rhythms of degassing, which occurs when the fluctuations in magmatic activity cause the lava lake to release increased concentrations and levels of gases into the air. It is a rhythm that allows the local population to feel the presence of the volcano and of the geological. In localities around Masaya volcano, PVEs are known to local populations as the hazy and blueish smoke carried by the wind, a mass of gas descending on the valleys and leading to skin rashes and respiratory disorders. This floating, toxic materiality provides PVEs with an ungraspable agency and a continuous presence for which the side effects are stretching the *now* into the *long-term*. On the one hand, their slow rhythms and erratic moves allow us to capture the interrelations between the otherness of geologic matter and the likeness of non-geological life, and on the other, between implicated and explicated temporal dimensions. Yet, this relationality is in turn posing important questions for the social studies of risk and disasters, as it is inviting us to reconsider what kind of time is privileged in risk management and how certain kinds of temporal awareness are excluded from the dynamics carried and used by DRR.

Central to DRR measures and response, resilience has become the concept that enables the mediation between instability and the capacity for human-centred life to reshape and to bounce back after shocks and threats. In the jargon of DRR, this mediating capacity is seen as the 'back to normality' effect, or what C. S. Holling (1996) defined as engineering resilience, which consists mainly of a capacity for a system to absorb a shock without causing major disturbances to its dynamics and functions, enabling the 'normal' state of that system to be sustained. Although there is a large array of literature that has critically engaged with the genealogy of resilience and that shows the multiple roots of the concept at the crossing of engineering, ecology, child psychology, neo-liberal economic theory and system theory (e.g. Alexander 2013; Chandler 2014; Davoudi 2012; Evans and Reid 2014; Grove 2014, 2018; Simon and Randalls 2016; Walker and Cooper 2011), the temporal underpinnings of resilience have remained less documented.

As argued elsewhere (Nobert et al. 2017), engineering resilience, as used and defined by DRR, became what Paul Virilio (1977, 2009) sees as an instrument of dromology: it embodies the valued speed and acceleration that are both essential to the realisation of the modern project (Conrad 1999; Eriksen 2001; Rosa 2010). The concept of resilience thus does not only radicalise bare life in contemporary politics and risk management efforts (see Agamben 1998). Nor does it solely aim at immunising neo-liberal life (Neyrat 2008; Esposito 2011). It also maintains a temporal relation: an appeal to move on quickly from adversities and difficulties so that one can reunite with what defines normality. This normality can then be reached by allowing us to keep our attention on what should come next – and, in doing so, resilience is stretching out the expected/known in time flow. Resilience allows us to focus our interest on the development of the *not-yet*, which is a temporal horizon that is not completely open as it is still entwined with the immediate past that fills our memory and provides the comfort of repetition, of the expected, of a future with few surprises (Bergson 1907, 1939).

The framing of time through the study of hazards

The lack of conceptual engagement with the kind of time resilience discourses embody and convey is not limited to the realm of DRR proponents; it also has its roots in the wider social sciences and humanities that have contributed to the development of disaster studies since the 1980s. A large part of this work has engaged with a conceptualisation of time as explicated, hinting at times at its implicated modalities and expressions, but without really acknowledging it as such. There are many epistemological reasons for the prevalence of explicated-uni-dimensional-clock-time in disaster studies, and aside from the dominance of the technical disciplines such as volcanology, geology or hydrology, social theory has also played an implicit role in maintaining a representation of time as the arrow at the bottom of the graph. One of those reasons can be found in the development of

Chicago's Hazard School and the prominent place John Dewey's pragmatism has occupied in its intellectual development, especially with his concept of the precariousness of existence and stability (Dewey 1929). Although Dewey's (1988) conceptualisation of time as an intrinsic value of human experiences is much closer to Henri Maldiney's (2012) definition of implicated time detailed above, the functionalism of prominent Chicago School hazards studies scholars such as Gilbert White, Robert Kates and Ian Burton has effectively marshalled Dewey's thought to produce time as an explicated variable within risk management initiatives that strive to control the future through techniques such as costs-benefit analysis and precautionary actions.

This erasure of the Deweyan pragmatist take on time is made more explicit with the interests in system theory and the quantitative turn that impacted on the development of American sociology that in turn helped reconfigure the field around metrology. Those changes have formalised an interest in numbers and a taste for rationality that had significant influences on the ways in which time has been accounted for by the heirs of the Hazard School. Perhaps more eloquently, though, it is the reliance on Herbert Simon's (1957, 1996) theory of choice that has played a significant role in erasing the Deweyan reference to the temporal processes involved in shaping the individuality of both organic and inorganic life (Dewey 1981, 1988). Drawing on social psychology and administrative science, Simon's theory mobilised a segmented vision of time that helped in understanding the functionality and dysfunctionality of the decision-making processes according to different states/stages of cognition: attention, emotion, habit and memory (Mileti 1999). The segmented cognitive process used by Simon and later replicated by DRR research refers to time as unidimensional, which in turn has made it easier for functionalism and human ecology to remain the main theoretical framings within which resilience has developed while continuing to orient the fields of risk and disaster management around the need for continuity, for which the rhythms of impulse and acceleration remained dominant (see Nobert et al. 2017).

Although Marxist political ecologists of the 1970s and 1980s have been highly critical of the Hazard School for disregarding the significance of political economy in the fabric of hazards and risks (see Walker 1979; Hewitt 1983), the temporal premises conveyed by the school and interrelated concepts of adjustments, adaptation and resilience were not central to their criticisms. Instead, they have been essentially drawing on Marxism and the calculative rationality of clock-time (i.e. the divided expressions of explicated time into units of seconds, minutes, hours and so on), which provided them with the necessary framing to explore and criticise the historic processes giving place to both commodification and compression of time in the production of vulnerability (e.g. Watts 1983) and later on in resilience (Joseph 2018).

This emphasis on the future and explicated time has not been limited to Marxist scholars. In spite of theoretical innovations in the ways in which time and movement have been addressed in social theory by poststructuralists, the engagement with risk, disasters and resilience has been mainly looking at the relational effects of power in the wider politics of life through the lens of a Foucauldian biopolitics (e. g. Keller 2015). If part of this work is rather critical of resilience as developed and used by DRR scholars and climate adaptation apostles, the emergence of assemblages and relational ontologies depict resilience as a *dispositif* (Braun 2013) capable of reorganising politics in a moment of crisis such as the one posed by the complexity of the Anthropocene. This reading of resilience is strongly associated with the geophilosophy of Deleuze and Guattari (1991: 45), which consists of not losing sight of the infinite (and complex) movement that allows thought to be creative (what they define as the re-territorialisation of thought), rather than seeking to provide reference points delimiting and restraining this movement to think anew. In both cases, the future remains the site of renewal, a performative category in which acceleration and high speed define the rhythms of change and dictate the transformation needed in our affective, political and relational dimensions of life in the face of the Anthropocene.

Bergson and the neglected possibilities of implicated time

Without entering into the details of Henri Bergson's vast philosophical project (i.e. thinking differences of nature outside the negation of dialectics) and complex terminologies, his work (i.e. Bergson 1907, 1927, 1939, 1941) has made it possible to move the concept of duration from the realms of mathematics and metrics to the centre of philosophical enquiries that can serve our reflection on the Anthropocene and resilience. Bergson's conception of duration fuses the past and the present together, making the present the most contracted level of the past and leaving the future to a state of actualisation of the extended past that is seen as being in perpetual movement, as invention (Bergson 1907; Deleuze 1966; Grosz 2004). This conception of duration is in turn helping us making sense of the contraction and extension processes that occur in the experience of time and that guide our attention and reorient our perception of what counts as time and, in the case of resilience, of the future.

It is important to insist that the experience of time through duration is not predetermined by a set of relations or rhythms. Rather, we must consider durations as linked to Henri Maldiney's (2012) conception of the 'real', which is the *unexpected*, the surprise that distorts explicated categories of time. Thus the state of 'normality' towards which resilience should lead us is disrupted by what Ilya Prigogine (1997) sees as the unexpected shifts in time continuum. Those shifts are eruptions from continuity that offer *transpossibilities*, that is, the possibilities to transcend the impossibilities established by the limits of positivism (Maldiney 1991: 75). In the case of the Anthropocene, an era that is calling for new relations with the non-human, an attention to the temporal relations emerging from rhythms and in the contracting and dilating processes of durations could help with understanding the transpossiblities offered by the unexpected brought by human relationships with geological life (see also Clark et al. 2018).

Locating the transpossibilities of rhythms and durations in the shaping of resilience to PVEs

Resilience, synchronisation and 'bouncing back' into the expected

Following the 2015 Third United Nations (UN) World Conference on Disaster Risk Reduction (WCDRR), the United Nations Office for Disaster Risk Reduction published the *Sendai Report: Managing Disaster Risk for a Resilient Future*, whose aims are defined as: '[t]he substantial reduction of disaster risk and losses in lives, livelihoods and health and in the economic, physical, social, cultural and environmental assets of persons, businesses, communities and countries' (UNISDR 2015: xi). Although most of the risks identified by the Sendai report are related to disasters, its expressed aims help us to understand what kind of time is defined and fostered by global efforts invested in building resilience in the face of the Anthropocene, emphasising that:

> Disasters have demonstrated that the recovery, rehabilitation and reconstruction phase, which needs to be prepared ahead of a disaster, is a critical opportunity to 'Build Back Better', including through integrating disaster risk reduction into development measures, making nations and communities resilient to disasters.
>
> *(UNISDR 2015: 21)*

Not only does this definition articulate the necessity to increase resilience to hazards and risks globally, but it also shows that in the expression 'build back better', the word 'back' indicates a predefined understanding of what 'normality' consists of and becomes the final destination that can be reached by speeding up the now ahead of the possible occurrence of a disaster to emerge. This urgency to secure the future is made possible by the quality of resilience, which in DRR means the capacity to leap forward, or a sort of conceptual trampoline allowing society to jump back to a state of expected outcomes. If the Sendai report is there to help with forging guidelines for a global strategy to build resilience in the face of upcoming risks associated with the Anthropocene era (e.g. climate change), those guidelines impose the rhythm of urgency and speed in relation to risks and to the need to realise and implement the UN Sustainable Development Goals as main political objectives. This fast forward rhythm imposed by the unexpectedness of risks and hazards has also led to the creation of the UN Plan for Disaster Risk Reduction for Resilience which aims to strengthen resilience in ways that:

> continue to cover the risk [from] disasters caused by natural hazards (geophysical, meteorological, hydrological and climatological) including extreme climate events, both slow and sudden onset, as well as strengthen activities related to other hazards, including technological and biological threats.
>
> *(UNISDR 2017: 9)*

Among those slow onset hazards for which the UN sees resilience needs to be developed, we find air pollution. In 2016, the World Health Organization (WHO 2018: 10) estimated that more than four million people died from illness related to outdoor air pollution, making it a silent killer that needs to be better tackled by the international community. Through the Sustainable Development Goals, resilience is seen as playing a major role in helping the world's population cope with air pollution, especially in urban centres and in the context of climate change. For example, the WHO (2017) has published a report entitled *Don't Pollute My Future! The Impact of the Environment on Children's Health* in which resilience to air pollution becomes seen as a way to react and to build intergenerational futures through the precipitated rhythm of urgency so that the future becomes composed of the expected, rather than shaped by the unexpected. In addition, there is a battery of so-called indicators that help identify what kind of risk management measures (e.g. structural and non-structural) should be taken to enhance the capacity of communities to bounce back to the expected – Westernised – futures in the face of disruptions.

In the case of Nicaragua for example, those processes are appearing in the framing of the strategic plan for the Pan American Health Organization in which there is a strong emphasis on:

> build[ing] the capacity of countries to protect the physical, mental, and social well-being of their populations and recover[ing] *rapidly* from emergencies and disasters. This requires adequate national leadership and sustained capacity of the health sector to build the resilience of countries and territories.
>
> *(PAHO-WHO 2014: 93; emphasis added)*

Resilience to air pollution and natural hazards is seen here as a must, and in this case, it is the fast forward quality of resilience that should be developed to reinforce institutions in the face of air pollution. There is thus an interest in imagining the futures of the region in ways that resonate with the explicated dimensions of time (i.e. past, present, futures), leaving the other temporal experiences and rhythms that are producing and that are produced by air pollution completely absent. This is particularly true for the air pollution resulting from geopowers that often resonates with a time felt and understood as multiple and as elastic, a time of surprise shaped by the unexpected and that disrupts predeterminations.

Meeting implicated time through the distortions of rhythms and durations

Asked why Theresa (fictive name) has decided to live in the village of Panamá, situated on the flank of Masaya volcano, she responded that,

> The smoking hill was there before I was born, [and] after the revolution those lands were redistributed or resold and some families managed to get them or

like in my case, my family bought it for me, [but] the smoky hill has always been there, there is not much I can do about it.

If this answer might seem trivial at first sight, it indicates a relation to the volcano that acknowledges deep geological time and the continuous presence felt through the 'smoke' (PVEs) coming from the 'hill' (Masaya). This answer also shows that the capacity to live in the village of Panamá, where inhabitants suffer from the effects of volcanic degassing, is about accepting the uncontrollable rhythms of geopowers. Theresa's relation to Masaya volcano was not unusual (as many people met through this research were accepting of the long-term presence and effects of the volcano), indicating that the fast forward clock-time futures encapsulated by the 'bounce back' rhythms of DRR resilience (and promoted as a 'solution' to health hazards) do not sit easily with the stretching effects of implicated time.

What consists of the now feels extended by the stretching effects of the ever-lasting geological rhythms of the volcano. If the extension of the now relates us to the experience of implicated time that reshape explicated clock-time experiences, what Martin Heidegger (1986) calls the *Unfung*, which is the feeling of a persisting actual that distorts its own transitivity in time flow, a sort of perpetual moment in temporal duration (see Nobert and Pelling 2017: 123) was also found among our interviewees. This *Unfung* becomes perceptible in the feeling of waiting for changes in the situation faced by Panamá's villagers, where the moment that consists of the now stretches and defies explicated time expression. For example, as noted during focus groups with villagers and civil protection, it became clear that children suffering from asthma as well as many adults experiencing exhaustion as a result of PVE pollution peaks were made worse according to the speed at which the winds carry the gases to the village. In those cases, villagers 'are staying inside their homes, waiting for the gases to pass and for the toxicity to decrease. But waiting for the winds to change direction might take hours that feel like a day' (interviewee 4, focus group 1). In this case, wind speeds are not only reinforcing the presence of geopowers in the lives of villagers, but they are also pushing people into their houses and in the process are extending the event that consists of the *now* into the unexpectedness of waiting. It is through this waiting that villagers 'meet' the implicated dimension of time in the shape of durations. The time felt by those impacted by PVEs seems to extend, bringing the implicated experiences of duration at the centre of the unexpectedness that fills the connotation of the not-yet and that makes the now feeling a-metrological. This stretched impression of the now is not only the result of waiting for PVEs to vanish, but also part of waiting for the only nebuliser available in the village to be borrowed and to help asthmatic children when PVEs are bad, waiting for governmental help to access fresh water, waiting to feel a priority for the Sandinistas, or waiting for political change (field notes 2017). These examples of temporal distortions are not the experience of the leap into the shiny future proposed by DRR functionalists' resilience, rather, we are in the slow rhythm of waiting that actualises what consists of the now and that brings upon us the surprises of the unexpected.

Villagers mentioned that when the rain starts, especially in the rainy season of winter, 'their skin feels different [and] that their eyes start burning' (interviewee 3), forcing them to develop several 'tactics' to deal with these side effects of PVEs. At the contact with the floating PVE plume, the rain becomes 'more acid, harsher on our skin, both adults' and children's and on vegetation' (interviewee 6). Rather than turning themselves to techniques and methods that would be dictated by public health authorities or international NGOs, many villagers mentioned during an information session (organised by the project and the national civil protection agency in February 2017) that they 'put plastic sheets over their water tanks' so that they can protect their skin from the acidic water when they are using this water to wash themselves (field notes, Panamá event 2017). Rather than establishing a protocol of practices to deal with PVEs, villagers are finding their inspirations in 'improvised' and in some cases experimental tactics, made in the flow of the now/moment and of surprises brought by the unexpectedness of discoveries. From the use of easy to reach plastic sheets by a mother to avoid the acidic water irritating her child's skin to the adolescent putting a 'piece of cloth of over his head in order to avoid the effect of PVEs on his eyes' (interview 5), the temporalities of those actions are not dominated by future planning, but are rather the result of serendipity, or the chance occurring at a specific moment. Although the continuity of life is central to them, they emerge out of the instant, of the unplanned surprises. Those practices exceed the unidimensional framings of DRR resilience and direct us to the unexpected possibilities that emerge and shape daily life; these possibilities for improvisation in the moment in the face of hazards are often overlooked when the focus is upon the realisation of the well-defined and linear future promoted through DRR resilience planning (e.g. UNISDR 2017).

The transpossibilities of the unexpected come from the blurring effects of implicated time, the time felt and experienced that occurs when 'solutions' to problems are part of improvisation and experiments. It became clear through the research process that since the 1970s and more particularly since the 1990s (focus group 1), degassing events became more frequent and new kinds of tactics and ideas become ways to deal with PVEs. For example, as seen in Figure 5.1, villagers have managed to find different ways to help protect their houses from the corrosive effects of PVEs, through the use of material found around them in the moment. These include stones, pieces of wood, plastic bags or ropes that are not corroded by the effect of PVEs. When villagers were asked how those materials were selected to improve their houses, most of them answered that they looked for material found near their dwellings (e.g. stones, wood). If this selection of material reveals the political economy found in Panamá, it also shows how the long-term predetermined futures of DRR resilience become disrupted by the possibilities offered by the unexpectedness of the elongated now that contracts the relevance of explicate futures, which here again operates through the logic one interviewee expressed: 'I took nearby rocks and branches that I found and used them to secure the roof tiles' (interview 9). If we are not in the sophisticated organisations of long-

FIGURE 5.1 Typical Panamá house on which the owners have used stones and branches to hold down the roof tiles (Photo taken by Xochilt Hernández Leiva)

term planning, the use of these materials means they can be replaced at any time and at low costs. This in turn is showing again a way of living that prioritises the actualisation of the moment rather than the configuration of univocal futures.

Conclusions

Although the time of resilience found in Panamá comes from a lack of political alternatives, it is important to understand that the fast forward DRR explicated futures are occurring simultaneously with other temporal experiences that have been overshadowed by the bouncing back to normality pushed forward by DRR proponents. It is here, between the capacity to embrace the unexpected through the distorting experiences of implicated time such as durations and rhythms and the need to consolidate expected futures, through urgency and speed, that the micro-politics of time-making takes place. This chapter has shown that while the ambition to secure the future through resilience is key to international discourses, what kind of time this future refers to has been widely unquestioned and taken for granted, leaving different ways of experiencing and producing time being ignored. The consequences for this omission might be more important than expected, mainly because what is conceived as an urgency to act, through DRR resilience mechanisms, might not correspond to the temporalities and temporal experience those, who are meant to be protected from PVEs' effects, can relate to. In the violence of the inequalities brought by a political economy of hazards, one can find an acceptance of the deep geological time of the volcano as well as of the implicated temporal dimensions that give place to multiple rhythms and durations of the everyday experience of PVEs.

The dominance of the explicated temporal realm in defining the essence of the future has somehow predetermined the political. It has prevented questioning of the implicated processes involved in actualising the experience of the now, which, as we have seen in the case of Panamá, provides the impulse to redefine what being-with others and the geological fabric of our planet means in the unexpectedness of implicated temporal processes. As we have demonstrated, being open to the trans-possibilites offered by the unexpected is in turn redefining the relationship between geopowers and human life, an important lesson in the era of the Anthropocene, which is well articulated through the effects of PVEs. In Panamá, nails, iron roof tiles and everything made of iron corrodes or becomes rusty in under a month under the transformative effects PVEs. Described by one interviewee as the 'smoke serpent', PVEs are dissolving minerals, turning them into rust and dust, bringing them back to the soil and thus to their point of origin, the deposits of the geological fabric of life. This transformative process offers a glimpse of the collision between the everlasting geologic life of Masaya volcano and the now of surprises serving to deal with the presence of PVEs. It is through this temporal gymnastic that we can see the shrinking of the now felt by villagers who fight to breath properly, who find solutions to growing food in those acidic soils and who protect drinking water from PVEs' pollution. Through this analytic we can understand that resilience is not a capacity to jump back to the normal. Rather, it is a play of rhythms, movement and durations that we have forgotten to question, but yet, are fundamental to life and to the new experiences and surprises that link us to the geological processes of our planet.

Note

1   This research was part of a project funded by a UK NERC-ESRC-AHRC Global Challenge Research Fund. The authors would like to thank the UNRESP team as a whole, but special thanks go to Evgenia Ilyinskaya, Peter Baxter, Caroline Williams, Hilary Francis and Tamsin Mather. All views in this chapter are those of the authors only.

References

Agamben, G. (1998) Homo Sacer: Sovereign Power and Bare Life. Stanford University Press: Stanford.

Alexander, D. (2013) Resilience and disaster risk reduction: an etymological journey. Natural Hazards and Earth System Sciences Discussions 1: 1257–1284.

Baxter, P. J., Ing, R., Falk, H. and Plikaytis, B. (1983) Mount St. Helens eruptions: the acute respiratory effects of volcanic ash in a north American community. Archives of Environmental Health 38(3): 138–143.

Baxter, P., Stoiber, R. and Williams, S. (1982) Volcanic gases and health: Masaya volcano, Nicaragua. The Lancet 320(8290): 150–151.

Bergson, H. (1907) L'évolution créatrice. Paris: Presses universitaires de France.

Bergson, H. (1927) *Essai sur les données immédiates de la conscience*. Paris: Presses universitaires de France.

Bergson, H. (1939) *Matière et Mémoire: Essai sur la relation du corps à l'esprit*. Paris: Presses universitaires de France.

Bergson, H. (1941) *La pensée et le mouvant*. Paris: Presses universitaires de France.

Bonneuil, C. and Fressoz, J-B. (2012) *L'événement Anthropocène: La Terre, l'histoire et nous*. Paris: Seuil.

Braun, B. P. (2013) A new urban dispositif? Governing life in an age of climate change. *Environmental and Planning D: Society and Space* 32: 49–64.

Chandler, D. (2014) *Resilience: The governance of complexity*. Abingdon: Routledge.

Clark, N. H., Gormally, A. M. and Tuffen, H. (2018) Speculative volcanology: time, becoming and violence in encounters with magma. *Environmental Humanities* 10: 273–294.

Conrad, P. (1999) *Modern Times and Modern Places: How Life and Art were Transformed in a Century of Revolution, Innovation and Radical Change*. New York: Alfred A. Knopf.

Davoudi, S. (2012) Resilience: A bridging concept or a dead end? *Planning Theory & Practice* 13: 299–333.

Deleuze, G. (1966) *Le Bergsonisme*. Paris: Presses Universitaires de France.

Deleuze, G. and Guattari, F. (1991) *Qu'est-ce que la Philosophie?* Paris: Les éditions de Minuit.

Delmelle, P., Stix, J., Baxter, P., Garcia-Alvarez, J. and Barquero, J. (2002) Atmospheric dispersion, environmental effects and potential health hazard associated with the low-altitude gas plume of Masaya volcano, Nicaragua. *Bulletin of Volcanology* 64(6): 423–434.

Dewey, J.(1929) *The Quest for Certainty*. New York: Capricorn Books.

Dewey, J. (1981) Experience and nature. In J. A. Boydston (ed.), *The Later Works of John Dewey, Vol 1*. Carbondale, IL: Southern Illinois University Press, 92–118.

Dewey, J. (1988) Time and individuality. In J. A. Boydston (ed.), *Later Works of John Dewey, Vol. 14*. Carbondale, IL: Southern Illinois University Press, 108–138.

Dupuy, J.-P. (2002) *Pour un catastrophisme éclairé: quand l'impossible est certain*. Paris: Seuil.

Eriksen, T. H. (2001) *Tyranny of the Moment: Fast and Slow Time in the Information Age*. London: Pluto Press.

Esposito, R. (2011) *Immunitas: The Protection and Negation of Life*. Cambridge: Polity Press.

Evans, B. and Reid, J. (2014) *Resilient Life: The Art of Living Dangerously*. Cambridge: Polity Press.

Grosz, E. (2004) *The Nick of Time: Politics, Evolution and the Untimely*. Durham, NC.: Duke University Press.

Grove, K. (2014) Agency, affect, and the immunological politics of disaster resilience. *Environment and Planning D: Society and Space* 32: 240–256.

Grove, K. (2018) *Resilience*. Abingdon: Routledge.

Hamilton, C., Bonneuil, C. and Gemenne, F. (2015) *The Anthropocene and the Global Environmental Crisis: Rethinking Modernity in a New Epoch*. Abingdon: Routledge.

Heidegger, M. (1986) [1950] *Les chemins qui ne mènent nulle part*. Paris: Gallimard.

Hewitt, K. (ed.) (1983) *Interpretations of Calamity from the Viewpoint of Human Ecology*. Boston: Allen & Unwin.

Holling, C. S. (1996) Engineering resilience versus ecological resilience. In D. P. C. Schulze (ed.), *Engineering Within Ecological Constraints*. Washington, DC: National Academy Press, 31–44.

Joseph, J. (2018) *Varieties of Resilience: Studies in Governmentality*. Cambridge: Cambridge University Press.

Keller, R. C. (2015) *Fatal Isolation: The Devastating Paris Heat Wave of 2003*. Chicago: Chicago University Press.

Lakhani, N. (2018) Nicaragua used 'weapons of war' to kill protesters, says Amnesty International. *The Guardian*, 18 October.

Latour, B. (2014) L'Anthropocène et la destruction de l'image du globe. In É. Hache (ed.), *De l'univers clos au monde infini*. Paris: Dehors.

Latour, B. (2015) Telling friends from foes in the time of the Anthropocene. In C. Hamilton, C. Bonneuil and F. Gemenne (eds), *The Anthropocene and the Global Environmental Crisis: Rethinking Modernity in a New Epoch*. Abingdon: Routledge, 145–156.

Maldiney, H. (1991) *Penser l'homme et la folie*. Grenoble: Editions Jérôme Millon.

Maldiney, H. (2012) *Regard, Parole, Espace: Œuvres Philosophiques*. Paris: Les éditions du cerf.

Mileti, D. S.(1999) *Disasters by Design: A Reassessment of Natural Hazards in the United States*. Washington DC: John Henry Press.

Neyrat, F. (2008) *Biopolitiques des catastrophes*. Paris: MF: Collection Dehorsm.

Nobert, S. and Pelling M. (2017) What can adaptation to climate-related hazards tell us about the politics of time making? Exploring durations and temporal disjunctures through the 2013 London heat wave. *Geoforum* 85: 122–130.

Nobert, S., Rebotier, J., Vallette, C., Bouisset, C. and Clarimont, S. (2017) Resilience for the Anthropocene? Shedding light on the forgotten temporalities shaping post-crisis management in the French Sud Ouest. *Resilience: International Policies, Practices and Discourses* 5: 145–160.

PAHO-WHO (Pan American Health Organization and World Health Organization) (2014) *Strategic Plan of the Pan American Health Organization 2014–2019*. Official Document No. 345. https://www.paho.org/hq/dmdocuments/2017/paho-strategic-plan-eng-2014-2019.pdf

Phillips, T. (2019) Nicaragua closer to new civil war than ever before, judge warns. *The Guardian*, 11 December.

Prigogine, I. (1997) *The End of Certainty: Time, Chaos and the New Laws of Nature*. New York: Free Press.

Rosa, H. (2010) *Alienation and Acceleration: Towards a Critical Theory of Late-Modern Temporality*. København: NSU Press.

Rosa, H. (2013) *Social Acceleration: A New Theory of Modernity*. New York: Columbia University Press.

Simon, H. A. (1957) *Models of Man*. New York: Wiley.

Simon, H. A. (1996) *The Sciences of the Artificial* (Third Edition). Cambridge, MA: MIT Press.

Simon, S. and Randalls, S. (2016) Geography, ontological politics and the resilient future. *Dialogues in Human Geography* 6: 3–18.

Stengers, I. (2009) *Au temps des catastrophes: Résister à la barbarie qui vient*. Paris: La Découverte.

Stengers, I. (2015) Accepting the reality of Gaia: a fundamental shift? In C. Hamilton, C. Bonneuil and F. Gemenne (eds) *The Anthropocene and the Global Environmental Crisis: Rethinking Modernity in a New Epoch*. Abingdon: Routledge, 134–144.

UNISDR (The United Nations Office for Disaster Risk Reduction) (2015) *Sendai Report: Managing Disaster Risk for a Resilient Future*. Geneva: UNISDR.

UNISDR (The United Nations Office for Disaster Risk Reduction) (2017) *United Nations Plan of Action on Disaster Risk Reduction for Resilience: Towards a Risk-informed and Integrated Approach to Sustainable Development*. Geneva: UNISDR.

van Manen, S.(2014)Perception of a chronic volcanic hazard: persistent degassing at Masaya volcano, Nicaragua. *Journal of Applied Volcanology* 3: 9. Virilio, P. (1977) *Vitesse et politique*. Paris: Galilée.

Virilio, P. (2009) *Le futurisme de l'instant: Stop-Eject*. Paris: Galilée.

Walker, J. and Cooper, M. (2011) Genealogies of resilience: from systems ecology to the political economy of crisis adaptation. *Security Dialogue* 42: 143–160.

Walker, R.(1979) Review of *The environment as hazard*. *Geographical Review* 69: 113–114.

Watts, M. (1983) On the poverty of theory: natural hazards research in context. In K. Hewitt (ed.), *Interpretations of Calamity from the Viewpoint of Human Ecology*. London: Allen and Unwin, 231–262.

WHO (2017) *Don't Pollute My Future! The Impact of The Environment on Children's Health*. Geneva: World Health Organization.

WHO (2018) *World Health Statistics 2018: Monitoring Health for the Sustainable Development Goals*. Geneva: World Health Organization.

Yusoff, K. (2013) Geologic life: prehistory, climate, futures in the Anthropocene. *Environment and Planning D: Society and Space* 31(5): 779–795.

Yusoff, K., Grosz, E. A., Clark, N., Soldanha, A. and Nash, C. (2012) Geopower: a panel on Elizabeth Grosz's *Chaos, Territory, Art: Deleuze and the Framing of the Earth. Environment and Planning D: Society and Space* 30: 971–988.

6

RESILIENT ARTS OF GOVERNMENT

The birth of a 'systems-cybernetic governmentality'

Sara Nelson

Introduction

This chapter returns to C. S. Holling's early resilience research at the International Institute of Applied Systems Analysis (IIASA) to show how ecological resilience theory was embedded in the production of new arts of government from its inception, with broad influence on the field of systems analysis much earlier than its mainstreaming in the 1990s. Drawing on Michel Foucault's comments on American neoliberalism, I argue that resilience inaugurates a mode of government that is environmental in two senses: first, because it intervenes in the conditions of life, and second because it instrumentalizes ecological dynamics to secure the life of populations. But this does not signal an inherent 'fit' between resilience and neoliberalism, in theory or practice. At IIASA, resilience gained influence in an interdisciplinary context that spanned the iron curtain, in which the relation among markets, state planning, and decentralization was an open problematic. This chapter shows how Holling's ideas influenced the way that systems analysis took shape in new arts of government that, while formative of neoliberal governmentality, are not defined by it.

As a concept of nature and a methodology of knowledge production, ecological resilience theory is fundamental to contemporary global change science. Even beyond influential institutions explicitly focused on the topic – including the Resilience Alliance and the Stockholm Resilience Centre – the understanding of nonlinear ecological change that resilience theory helped to inaugurate has achieved the status of a new paradigm in the environmental sciences, and provides much of the basis for contemporary responses of climate change through theories of adaptability and transformation (Pelling 2011). This is true of the content of knowledge as well as the process and institutions of its production; as this chapter describes, the history of resilience involves new ways of generating knowledge in conditions of irreducible uncertainty (through gaming, scenarios, and adaptive management techniques), and new institutional forms for addressing problems in complex industrial societies (specifically, the international think tank). In important ways, it is through ecological resilience theory and its inheritors that we have come to know, and to respond to, the Anthropocene.

This chapter revisits the early development of resilience theory in ecologist C. S. Holling's work on resilience at IIASA, where he was head of the Ecology Program from 1973 to 1974 and later director from 1981 to 1984. I argue that ecological resilience was embedded in the production of new arts of government from its inception – what Egle Rindzeviciute (2016) has called a 'systems-cybernetic governmentality.' In this early work resilience was not strictly an ecological concept that was subsequently expanded to describe social problems, but was developed as a universal systems language that would address conjoined problems in ecology, economics, and social and institutional management, with broad influence on the field of systems analysis much earlier than its mainstreaming in the 1990s. Drawing on Foucault's comments on liberal biopolitics, I argue that resilience theory posited a new ontology of nature that challenged political economy as the principle of limitation internal to governmental reason, offering a liberal critique of modern resource management. I argue that resilience articulates an environmental art of government in a double sense: it operates on the conditions of life – reflecting Foucault's descriptions of American neoliberalism – and it also enrolls environmental processes in the government of populations. In this way resilience has been instrumentalized under neoliberalism as a mode of government-through-environment, shaping our experience of the Anthropocene as a generalized condition of environmental precarity.

However, drawing on Egle Rindzeviciute's (2016) history of systems science at IIASA, I argue that this environmental mode of government is not specific to neoliberalism; rather it was developed through international collaboration among scientists in socialist and capitalist countries. At IIASA, the concept of resilience was at the center of an international and interdisciplinary research program that spanned the iron curtain, in which the relation among decentralization, markets, and state planning was an open problematic. This history shows that resilient styles of government have no inherent 'fit' with neoliberalism; the challenges posed to liberal government by 'complex life' (Chandler 2014) were also challenges for state socialism, and resilience may have linked up differently with socialist political projects. In this way the chapter joins other work (Grove 2018, Chandler 2014, Schmidt 2015, Nelson 2014) demonstrating the counterrevolutionary origins and orientation of resilience, or the conflicted ways that resilience and related complexity sciences are taken up and strategically deployed within neoliberal forms of rule, while still retaining other lives.

I begin by historicizing resilience in the context of ecological and environmental concern in the 1960s and 1970s. I then focus on the early development of resilience theory at IIASA, showing its influence on new approaches to modeling complex problems and new understandings of and ways of responding to catastrophic risk. As one of IIASA's most successful and influential early endeavors,

resilience has not only shaped the way that we know and respond to environmental change, but has also helped to bring environmental management to the
forefront of governmental practice. The tools for the production of resilient forms
of life in the present were forged in this Cold War context, where the political
implications of resilience and related forms of systems research were not yet
determined. In this way the chapter gives historical detail to the assertion that other
articulations of resilient life are and were possible.

Historicizing resilience

In his late lectures, Foucault (2008: 21–2) argues that the emergence of biopolitics
cannot be understood without reflecting on its historical conditions in liberal governmental reason. Unlike the juridical critiques of state power based on natural
right, which called on transcendent principles, Foucault argues that liberalism is
defined by a principle of limitation to state power that is internal to the objectives
and rationality of government itself, based on a certain understanding of 'nature'
(Foucault 2008: 10, 20–1).[1] For Foucault, political economy is the 'intellectual
instrument, the form of calculation and rationality' that made this self-limitation
possible (2008: 13):

> [P]olitical economy does not discover natural rights that exist prior to the
> exercise of governmentality; it discovers a certain naturalness specific to the
> practice of government itself … The notion of nature will thus be transformed
> with the appearance of political economy. For political economy, nature is not
> an original or reserved region on which the exercise of power should not
> impinge, on pain of being illegitimate. Nature is something that runs under,
> through, and in the exercise of governmentality … [I]f there is a nature spe-
> cific to the objects and operations of governmentality, then the consequences
> of this is that governmental practice can only do what it has to do by
> respecting this nature.
>
> *(Foucault 2008: 15–16)*

Foucault argues that the market becomes a 'site of veridiction' within liberalism,
such that the measure of good government is not: is it legitimate? But rather: is it
effective? Foucault thus defines liberalism less as a coherent doctrine than a principle of critique of governmental practice, defined by the questions: Why govern?
and How does one not govern too much? 'The economy' was a crucial mechanism
through which the politics of population were mediated by the institution of the
market as a site of veridiction and test of government: a new object of scientific
knowledge and management that emerged between the 1930s and 1950s, the economy was conceptualized as a nationally-bounded entity with a normative tendency
toward unlimited growth (Mitchell 1998). As Mitchell (1998: 93) writes, 'To create
the economy meant also to create the non-economic.' While Mitchell focuses on
the devaluation of subsistence production and social reproduction, this also entailed

the exclusion of ecological reproduction. The environment was the constitutive outside of the economy in both a literal sense as the source of material inputs and repository for waste, and in a discursive sense as the field of material wealth and productive activity in contradistinction to which the economy's boundaries must continually be drawn.

Placing Mitchell's account in the context of the political economic disruptions of the 1960s and 1970s, however, we can observe that almost as soon as the economy came into being as a coherent object of government, it was thrown into crisis.[2] That period saw a proliferation of literature in which environmental issues, rather than localized events or problems linked to individual resources, came to be seen as part of a globally integrated set of complex systems whose dynamics defied pre-diction and whose biophysical limits threatened economic growth (e.g. Carson 1962, Meadows et al. 1972, Ehrlich 1968, Lovelock and Margulis 1974; see Nelson 2015). This literature articulated a crisis that was environmental not because it pertained strictly to nonhuman issues and populations, but because it was a crisis of the boundaries of the economy itself; the latter's constitutive blindness to its own ecological conditions was discovered to pose a series of threats to the health of the social body, including looming scarcities of critical resources and the unpredictable feedbacks generated by social-ecological complexity (Nelson 2015).

In this context, the debates that pitted 'technological optimists' like Robert Solow and Julian Simon against proponents of ecological limits like Herman Daly and Paul Ehrlich concerned the extent to which the 'truth' of the market was a sufficient measure of effective government, and to which liberalism, as a principle of limitation to governmental action, was compatible with an ecologically-attuned capitalism. Much like the principle of right that Foucault contrasts to the liberal critique of government, thinkers of natural limits posed the environment as an external limit to government action, based on a transcendent principle of ecological balance. But formulas for the no-growth society, while influential, were already out of touch with political realities. Instead, a very different type of response to these urgencies was taking shape, as Foucault indicates in his discussion of Amer-ican neoliberalism in the 1970s:

> [W]hat appears on the horizon of this kind of analysis is not at all the ideal or project of an exhaustively disciplinary society ... On the horizon of this ana-lysis we see instead the image, idea, or theme-program of a society in which there is an optimization of systems of difference, in which the field is left open to fluctuating processes, in which minority individuals and practices are toler-ated ... and finally in which there is an environmental type of intervention instead of the internal subjugation of individual.
>
> *(Foucault 2008: 259–60; see also Zebrowski 2014)*

Foucault describes the arts of government emerging in the work of American neoliberals like Gary Becker as 'environmental' in the sense that they work to influence individual action by shaping the milieu in which that action takes place.

The key shift for Foucault is from a disciplinary system seeking to cultivate internalized control through externally-imposed order, to an approach that grants the individual autonomy within a framework in which the effects of his or her actions can be modulated to encourage or discourage particular behaviors.[3] As Foucault writes, this is '[n]ot a standardizing, identificatory, hierarchical individualization, but an environmentalism open to unknowns and transversal phenomena' (Foucault 2008: note, 261). In other words, this type of government-through-environment (in the broadest sense) could tolerate a broad range of behaviors on the level of the individual with the understanding that this behavior displayed some range of systematic responsiveness to its environment.

As Chris Zebrowski (2014) has argued, Holling posed the environmental problem in terms of a new nature that demanded new forms of knowledge and management that would enable precisely an 'optimization of systems of difference' and an 'environmental type of intervention.' In place of an equilibrium ecology amenable to top-down control (the correlate to the 'normalizing society'), Holling posited a new ontology of nature characterized by a plenitude of ecological capacities that were being stifled by attempts to impose equilibrium as an external norm. In a seminal 1973 paper, he introduced resilience as an explicit counter to the emphasis on stability prevalent in notions of 'ecological balance' (Holling 1973a: 1). Based on his research into spruce budworm outbreaks in British Columbia,[4] Holling argued that systems have multiple equilibria or 'domains of attraction,' and may be highly resilient precisely because they are unstable. Moreover, modern resource management practices that sought to impose stability had the perverse effect of reducing system resilience. Through resilience, Holling proposed a style of management that intervened in the environmental perturbations driving system behavior, encouraging a range of play with ecological dynamics to foster their emergent normativity. In doing so, he posed a critique of liberal government on its own terms: rather than invoking ecological balance as a transcendent moral principle, resilience raised a question of effectivity — how should management work with the dynamics of 'complex life' (Chandler 2014) in order to most effectively foster and control its inherent capacities?

By deploying a liberal style of critique to push liberal government beyond itself, resilience shares with neoliberalism the function of 'updating key tenets of liberal thought' in light of novel situations (Grove 2018: 14; Collier 2011). But this does not mean that its neoliberal iterations were predetermined in its origins. In this regard Holling's tenure at IIASA becomes significant: at IIASA Holling developed concrete strategies and tools for governing this new nature — for working with its emergent normativity in the absence of total knowledge of the system and predictive power – in relations of mutual influence with systems analysis in fields ranging from economics to energy policy and urban design, and in ways that were relevant to both socialist and capitalist societies. Through his collaborations with scholars in management sciences, economics, and energy from both East and West, Holling's ideas influenced the way that systems analysis took shape as a new 'art of governance' (Rindzeviciute 2016: 9) in diverse political economic contexts.

Holling at IIASA

Founded in 1972, IIASA arose in the context of what I have called the 'environmental crisis' to facilitate collaboration between socialist and capitalist nations on problems of global scope and universal relevance to industrialized societies. The institute was the brainchild of Aurelio Peccei, the president of the Club of Rome and the major force behind the production and promotion of the 1972 Limits to Growth report, who envisioned the center as part of a broader strategy to inaugurate a new system of global governance (Peccei 1969).[5] As Rindzeviciute (2016) describes, IIASA was the first international think tank, establishing an influential model of knowledge production that would proliferate in the following decades. She argues that the institute was for-mative of a new 'systems-cybernetic governmentality' which 'constructed the world as a set of complex and dynamic systems, consisting of different geological, biological, and technical phenomena, which were also subject to tactical regulation in the same way as population was for Foucault'[6] (Rindzeviciute 2016: 8). Moreover, she shows that systems-cybernetic arts of government were co-produced at IIASA through col-laboration among socialist and capitalist countries, rather than simply being dis-seminated from Western scientists. These techniques 'transformed the very character of control' in Soviet societies, smuggling in 'a new, postpositivist notion of the human and nonhuman systems' under the guise of a depoliticized technology (2016: 13). In this way the story of IIASA is one of the movement from a first-generation science of closed equilibrium systems to second-generation systems science defined by multiple equilibria and resilience, and of the significance of this shift for new technologies of power that were deployed in diverse political economic contexts.

This chapter adds to Rindzeviciute's analysis by highlighting the importance of Holling's resilience research to the systems-cybernetic governmentality developed at IIASA. Holling joined IIASA's first cohort in 1973 as head of the Ecology Project, along with ecologist Bill Clark and systems engineer Dixon Jones from the University of British Columbia (UBC). Holling's initial proposal focused on the planning and policy implications of resilience, describing a new ontology of nature that had been misunderstood and undermined by previous management approaches:

> Up to now the resilience of these systems has allowed us to operate on the presumption of knowledge with the consequences of our ignorance being absorbed by the resilience. Now that the resilience has contracted, traditional approaches to planning might well generate unexpected consequences that are more frequent, more profound and more global. The resilience concept pro-vides a way to develop a planning framework that explicitly recognizes the area of our ignorance rather than the area of our knowledge.
>
> *(Holling 1973b: 3)*

The Ecology Project was to become one of the most productive and successful projects of IIASA's early years; Howard Raiffa (2011: 97), IIASA's first director and founder of Harvard's Kennedy School of Government, described it as 'one of our best.' The project attracted the participation of the top IIASA scientists in this early period, including Tjalling Koopmans from Yale, who would win the Nobel prize in economics during his IIASA tenure; George Dantzig from Stanford, the originator of the simplex algorithm and the father of linear programming, who led IIASA's Methodology Project – and who both Raiffa and Koopmans agreed should have shared Koopmans' Nobel (Raiffa 2011: 211; Mirowski 2002: 289); and Alan Manne, also from Stanford, a leading energy economist (Raiffa 2011: 112; Rindzeviciute 2016). As Raiffa (2011: 111) writes, Koopmans, Dantzig, and Manne 'for a year or so became budworm experts.'

Strengthening the interdisciplinary applications of resilience was a key focus of Holling's work at IIASA (Jones 1977). In a planning conference that first year, the links between resilience research and game theory, participatory modeling, urban design, energy systems, and water resources were already being forged (IIASA 1973). In his summary of the proceedings, Holling wrote that the study of complex ecosystems 'underlie[s] every other conceptual, methodological or applied project conceivable for IIASA': on the one hand new forms of ecological management were essential to addressing emergent global problems, and on the other ecosystems displayed universal qualities of complex systems, such that new strategies of ecological management bore direct relevance to other fields (Holling 1973c: 7).

That conference planted the seeds for much of the interdisciplinary work that would be pursued in the coming years.[7] A key point of consensus was the need to translate resilience from a descriptive model into new environmental indicators and management approaches that took account of non-linear dynamics (IIASA 1973: 28). Holling articulated this need in terms of a shift in environmentalism away from what he described as the 'eco-freak' model and its antagonistic stance toward business, including the then-recent ban on DDT and efforts to oppose freeways and dams (Holling 1974a: 1). In contrast, Holling described a new role for ecology in 'adjudicat[ing] some of the inevitable conflicts between economic needs and environmental (ecological) protection:'

> [W]e are not so naive as to crusade wholly in the name of generations yet unborn and to march off to battle against politicians and captains of industry. Our analysis is directed at those who would study ecological currents and forces, who would ride them piggy-back rather than subdue them to any end, however noble. It is directed at those who would undertake ecological engineering simply because it may be the most cost-effective engineering for certain laudable commercial enterprises, not merely because it is the gentlest, the most humane, the most natural, or (somehow) the most acceptable. Our objective is not merely to preserve ecosystems (whatever that means) because it is fashionable to do so, but rather to encourage the systems to work with us.
>
> *(Fiering and Holling 1974: 2)*

Holling framed the 'new science of ecological management/engineering' (Holling 1974a: 1) as a utilitarian one that would integrate economics and ecology in order to develop practical techniques for cultivating and harnessing ecological capacities. Rather than natural equilibrium imposing external constraints to the expansion of the economy (as in the work of ecological economists like Herman Daly), for Holling economics provided a type of rationality to be integrated into environmental decision-making. On the basis of innovations in applied ecology, mathematics, computer modeling, and management sciences – especially policy analysis and decision theory – ecology could progress beyond description to become a prescriptive science (Holling 1974a: 2–3).

Holling took the first steps toward this new science in collaboration with the engineer and mathematician Myron Fiering to develop new environmental standards and indicators of environmental quality based on resilience. In contrast to previous ecological models, Fiering and Holling (1974) defined indicators in terms of functional roles in the ecosystem rather than individual species, distilling complex ecosystem behaviors to their key elements in order to create a library of 'modules' that could replicate a range of possible system behaviors. Borrowing IIASA's terms, this shifted the focus of modeling from a global to a universal perspective, abandoning the possibility of a total picture of the complete system in favor of a streamlined catalog of universal types of system behavior which could be effectively modeled with a minimum number of variables.

This shift from species to roles opened up a broader description of 'normal' ecosystem behaviors, while also suggesting new prescriptive approaches for working with this normativity (or, as they put it, 'encourag[ing] the systems to work with us'). There was no reason to presuppose, they argued, that a previous system state was desirable; a passage beyond a resilience threshold into a new stability domain may be preferable, and might be deliberately accelerated 'by selective seeding or killing of organisms' (Fiering and Holling 1974: 9). Measures of resilience would enable managers to assess how cost effective it may be to maintain the system in any given state (1974: 18–20; 30). These modules would provide the basis for Holling's influential Adaptive Environmental Assessment and Management (1978) which, as Holling described, systematized the approach into 'packages of techniques […] available for broad applicability to any problem of single species management, whether of pests or crops or of man or of populations of fish or wildlife' (quoted in Hutter 2018: 271).

Resilience and adaptive management shifted the object of ecosystem management from individual species to the relational qualities and capacities of ecosystems to grow, change, and produce desired outcomes. Resilience intervenes in ecological relationality to work with ecosystems' emergent capacities and harness these toward management goals. As Isabel Schrickel (2014: 15, my translation) writes, at IIASA resilience is 'on the one hand discovered as an emergent property of ecosystems, [and] on the other hand can be understood as an environmental embedding of governmental practices in social and economic systems.'

The relevance of the ecology research to modeling and managing complexity in other fields was widely recognized by IIASA scholars. By addressing formal problems of unstable system behavior with multiple equilibria, Holling's work brought ecology to the forefront of advances in differential topology, catastrophe theory, and mathematical modeling. Mathematicians in IIASA's Methodology Program 'recognized the immediate connections' between Holling's phase portraits and the global theory of differential equations, while Koopmans noted the close resemblance between Holling's equilibrium points and new fixed-point algorithms in economics (Grümm 1977: 5). Holling's colleague at the IIASA, Bill Clark, recalled in a personal interview the collegial atmosphere that facilitated these connections:

> Koopmans was really interested in what we were doing. And you know he would pose a question, and then listen, and then pose another question, and so on. And he was clearly getting more and more interested in the notion of multiple stable attractors, what is now after Brian Arthur thought of as non-equilibrium economics ... Koopmans was out ahead of any of the rest of us in seeing how there was similar stuff happening across these multiple different systems and trying to get us together to begin to see if we could come to some framework that would let us use the same terminology and where appropriate the same analytics, so that we could begin to compare experiences and see what was sui generis and what was similar and so forth.
>
> *(Clark 2018)*

In 1975 Koopmans organized and chaired an international conference that explored the relevance of resilience theory to mathematical problems in the areas of meteorology, climatology, economics, and chemistry, among others (Grümm 1977; Grümm 1975). At the workshop, members of the Ecology Program worked in collaboration with scientists from Eastern and Western member countries on the formal problems presented by resilience dynamics. The workshop inspired further research in climate modeling and chemical evolution at the University of Vienna, MIT, and the Frei Universität Berlin. It also solidified the centrality of resilience research to other projects at IIASA, primarily the Energy and Methodology programs (Grümm 1977: 6). A 1976 workshop on pest management would further applications of resilience in Eastern bloc countries (Hutter 2018: 277).

These intersections also shaped resilience science. George Dantzig proposed dynamic programming as an alternative to the Ecology Group's successive simulation experiments, bringing 'the great insight that said that you could in fact use smart computation, not analytics, to work backwards from desired solutions to figure out optimal pathways to get there rather than working forward from possible branch points' (Clark 2018).[8] One of the strongest areas of methodological connection was between the Ecology Project and the Energy Project, the latter spearheaded by Wolf Häfele from the Karlsruhe Nuclear Research Center in West Germany and IIASA's Deputy Director (Clark 2018; Schrickel 2014). An early thinker of energy transition, Häfele was concerned with the prospect of climate change and saw in the ecology research a way of understanding 'how deeply we were in this deep attractor of fossil fuel systems' (Clark 2018). As the senior scholar most connected to the policy world, Häfele's deep interest legitimated the ecology work and helped to extend it beyond the environmental realm (2018). As Bill Clark (2018) recalled, 'he like Koopmans had no problem at all grabbing one of us juniors and saying let's go to lunch, talk to me about this, come for a ride in my car, which was invariably a broken down VW Beetle rather than the Mercedes that all the other heavyweights were driving around.'

The intersection with the energy project pushed resilience research in new directions. By delinking resilience from the conservative bias of ecology (its concern with preserving existing systems), it was Häfele who 'turn[ed] the problem upside down' to ask how systems get locked into undesirable domains of attraction, and how they might be deliberately pushed out (Clark 2018). Studies in this kind of social-ecological transformation now constitute a vibrant area of research at institutions like the Stockholm Resilience Centre. The intersections with the energy project also point to other descendents of resilience: according to Bill Clark, much of the current thinking on energy transitions like the German *Energiewende* is 'the direct linear descendent of Häfele's thinking as cross fertilized with the environment notions,' representing a more fruitful lineage of Holling's work than the subsequent forays into panarchy theory and Schumpeterian creative destruction.[9] Through these collaborations, resilience theory took shape in an interdisciplinary environment in which it helped to shape the tools currently used to understand and respond to the urgencies that define the Anthropocene.

As in other areas of IIASA, advances in resilience science were catalyzed by combining Soviet technical capacities in mathematics and modeling with conceptual advances from Western academies (Rindzeviciute 2016: 117). A major breakthrough for the Ecology Project came through collaboration with the Russian mathematician Alexandr Bazykin, who demonstrated the power of contemporary work in differential analysis during a lecture that Clark gave on the budworm simulation experiments:

> so I had begun to sketch what the state space of this system seemed to look like. And he said 'uh, well, that's very interesting, but couldn't you interpret the same system this way?' ... [and he] wrote down three coupled differential equations on the blackboard and solved them as we watched, into an analytically derived state space that was the same as our state space! ... [B]asically he opened our eyes to what applied mathematicians working in differential analysis were doing ... It was seriously sobering.
>
> *(Clark 2018)*

The key shift brought by Bazykin was the strategy of 'model[ing] the model,' or building differential equation models of the state space produced through the simulation experiments that could reveal key areas for further exploration with the more complex computational model (Clark 2018). These results then could be confirmed with field data. This intervention revolutionized the approach to modeling complex ecological systems, leading to a number of collaborations between Holling, Bazykin, Dixon Jones, and others. This way of working between multiple models validated by field data is now widely used in climate modeling, engineering, and policy analysis, and the work at IIASA was some of the first to demonstrate its practical application (Clark 2018).

At IIASA resilience research was also influenced by scholars from the global South. A 1974 workshop on global modeling focused on a project led by Gilberto Gallopín at the Bariloche Institute in Argentina. In direct response to the controversial Limits to Growth report, that project demonstrated that global crisis could be avoided by a global transition to a democratic socialism. Gallopín later participated in IIASA conferences on environmental impact assessment, and he and Holling organized a collaboration among IIASA and Bariloche on 'coping with surprise' during Holling's tenure as IIASA director in the early 1980s. In early exchanges about the project, Bariloche scholars challenged the implicit normative emphasis on resilience and argued for an emphasis on deliberate system change, advocating for a focus on aesthetic, social, and ecological values as well as funding arrangements to support the involvement of Third World researchers (Gallopín 1980). In his contribution to the project, Gallopín pushed the conversation on 'surprise' to contend with socio-political instability by using the institutional history of the Bariloche Institute in the context of political upheaval in Argentina in the 1970s as a case study (Gallopín 1981).

A defining feature of resilience, and a key point of intersection with other fields at IIASA, was a new way of understanding and dealing with uncertainty. For Holling, resilience was concerned with 'qualitative unknowns' stemming from non-deterministic nature itself: he wrote that it was impossible to simply 'engineer nature, (i.e. the unknown) out of the equation' (Holling 1974a: 4).[10] It was this topic of management under conditions of irreducible, non-probabilistic uncertainty that catalyzed many of resilience's interdisciplinary connections, in part through an informal 'resilience group' that collaborated on decision analysis and management sciences (Schrickel 2014: 16; Bell 1977). This was a particularly fruitful area of collaboration between the Ecology and Energy Projects. Häfele, a strong advocate of nuclear energy,[11] was preoccupied with the problem of 'hypotheticality' — the idea that socio-technological systems had developed to the point at which the prospect of catastrophic disaster had rendered trial-and-error approaches to engineering infeasible. As Holling and colleagues summarized, 'we are locked in a world of hypotheses because we dare not test our hypotheses' (Jones, Holling, and Peterman 1975: 3). Holling and Häfele immediately saw a resonance between the notions of hypothetically and resilience. As they wrote in an invitation to participants in a workshop on Resilience, Hypotheticality, and Option Foreclosure in 1975, 'Each concept, for instance, is centrally concerned with understanding the nature and implications of the *unknown* (as distinct from the merely uncertain) in various aspects of applied systems analysis,' and each 'breaks with trial-and-error as a sufficient paradigm for program development' (Holling, Häfele, and Walters 1975: 3, original emphasis).

Schrickel (2014) argues that Häfele latched onto resilience as a strategic discourse for selling nuclear energy in the midst of West Germany's powerful antinuclear movement, portraying atomic power as part of a resilient energy system, and that the topic was a point of continual tension between Holling and Häfele and within the resilience group. As summarized in the proceedings of the 1975 conference, Holling stated that the overall aim of his work was to design ' "a world without hypothetically": not to design yet more coping strategies that dug us all in deeper, but to prescribe approaches that would make the consideration of hypotheticality irrelevant altogether' (Clark 1975: 8). Nevertheless, Häfele would continue to invoke the resilience concept to argue against small-scale renewable energy over the following years (e.g. Häfele 1981), and they continued to collaborate with the aim of finding common ground among the concepts after Holling had returned to UBC (e.g. Häfele, Balinski, and Holling 1975).

Their discussions centered on what Holling saw as the necessary and generative role of crisis in resilient systems. At the 1975 conference on Resilience, Hypotheticality, and Option Foreclosure, Holling and colleagues argued against efforts to minimize the risk of system collapse, suggesting that such strategies 'favored, unwittingly, a move away from resilience and toward hypotheticality' (Swain 1975: 9). Offering the figure of the 'entrepreneur' as the one who 'need[s] risks, need[s] unexpected events for personal enrichment,' but who also gives 'high value' to such events 'almost irrespective of benefits,' they argued that the periodic experience of catastrophe was essential to the system's learning, such that 'there might well be a place in environmental, institutional or societal management for disaster design — periodic 'mini-disasters' that prevent the evolution of inflexibility' (Jones, Holling, and Peterman 1975: 4, 6). The emphasis on a productive relation to disaster and disturbance would be strengthened in Holling's subsequent work; in his 1978 edited volume Adaptive Environmental Assessment and Management, the first major output to synthesize a resilient management agenda, Holling (1978: 11; original emphasis) defined resilience as the capacity 'to absorb and *utilize* (or even benefit from) change.' This was further developed in his influential concept of the adaptive cycle, in which crisis becomes a moment of 'creative destruction' when system resilience is either renewed or destroyed to allow for a new configuration to emerge (Holling 2001).

The generative role of crisis in resilience theory is a key point of departure for many commentators, who have pointed out how resilience naturalizes the turbulence of late capitalism and makes vulnerability a fundamental condition of social life (Walker and Cooper 2011; MacKinnon and Derickson 2012; Evans and Reid 2014). Without invalidating these critiques, the trajectories of resilience at IIASA point to other possible iterations, while highlighting the disjunctures between resilience theory and neoliberal imperatives. An alternative vision of resilience appears through Holling's reflections on a visit to China that he took during his first year at IIASA, the same year he published his seminal paper on resilience. Holling described the communal structure of Chinese agriculture as a highly resilient system characterized by local self-sufficiency and decentralization that enabled responsive governance: 'Behind the cant and slogans,' he wrote, 'lies a brilliantly structured and humane strategy which emphasizes persistence not efficiency, equality not diversity, and self-sufficiency not dependence' (Holling 1973d: 2). In an essay entitled 'On Making a Marriage, an Institute, or a Society,' Holling used China's agricultural system as an example of resilient strategy relevant to governance problems within institutions of all kinds (including IIASA). Holling praised the internally diversified, locally self-sufficient communes and villages, which he noted violated received economic doctrines of comparative advantage, efficiency, and economies of scale (Holling 1973e). By prioritizing self-sufficiency at the small scale, the Chinese system, in Holling's eyes, was better able than the West to foster innovation and independence on the part of local units, while minimizing the costs of failure for society as a whole (Holling 1973e: 3–4). He contrasted this to British imperial policy in Burma, where privatization and market reforms combined with hierarchical governance created a maladaptive, inflexible system that could not respond to drought or other emergencies (1973e: 4). For Holling, this suggested lessons for IIASA's own leadership and for the application of systems analysis as a governance strategy:

> True strategic failure occurs when solutions lie in other, isolated hands. That is the route which leads to systems analysis as the panacea of the world's ills — glue for the unglued. The successful strategy then makes possible the easy exploration of alternate tactics. One important set are the tactics of initiating a fresh concept and direction within a system which by definition resists change. And these tactics must dance between the extremes of dictatorial fiat and subtle deviousness, both of which are seeds for disaster.
>
> *(Holling 1973e: 5)*

In other words, resilience for Holling was not a management directive to be imposed from the outside but a strategy of experimentation emphasizing localized control and functional redundancy in ways that do not necessary accord with doctrines of efficiency and profit maximization. Holling's early work at IIASA indicates not only the concept's influence on multiple fields, but also that resilient forms of life might take shape in a diversity of political economic contexts. This turns attention to the tensions and inconsistencies that emerge as notions of resilience are tacked onto neoliberal policy imperatives.[12]

Conclusion

In his reflection on IIASA's first five years, Roger Levien, IIASA's second director, summarized the importance of the Ecology Project:

> the approach developed and the analytical issues addressed turned out to have relevance far beyond the specific subject matter of their application ... Holling and his colleagues addressed complexities that systems analysts generally ignore: multiple conflicting decisionmakers; multiple conflicting objectives; inter-temporal and inter-generational trade-offs; and the design of strategies that deal with the irreducible uncertainty of the real world by, for example,

avoiding premature foreclosure of options or by being 'resilient' ... These ideas, spawned in the Ecology Project, influenced other projects as well, particularly Energy and Methodology.

(Quoted in Schrickel 2014: 17)

The work conducted by Holling's early team set the direction for subsequent research at IIASA and was integrated into domestic environmental management programs in socialist and capitalist countries (Hutter 2018). East-West collaboration on environmental problems at IIASA laid the groundwork for international research on climate change (Rindzeviciute 2016). But less well-acknowledged is its significance to early advances in systems analysis and its governance applications more broadly. Through his collaborations with Häfele, Holling's work shaped new understandings of energy systems and new strategies for contending with risk (McDonald, no date; Clark 2018). Holling and colleagues' observations of multiple equilibrium dynamics helped to demonstrate the relevance of advances in topology and differential equations for similar problems in economics, climatology, chemistry, and other fields, and brought ecology to the cutting edge of complex systems modeling. In economics, many of these ideas would come to fruition much later through the complexity economics of Brian Arthur, another alum of the early IIASA years (Clark 2018; Arthur 2014).

The advances in decision analysis and computer modeling in the Ecology Project helped to pave the way for new applications of systems analysis in planning and management in situations of high complexity, while their experimental workshops for adaptive management pioneered new methods for knowledge production about unknowable futures, including interactive gaming and early scenario planning enabled by new computer models (Schrickel 2014; Hutter 2018).[13] These strategies have evolved through the further development of adaptive management, and gaming remains an important strategy for cultivating 'resilience thinking' through initiatives such as 100 Resilient Cities (van der Meyden 2014; Kupers, personal interview, March 15, 2016). Holling's experience at IIASA demonstrates that resilience, rather than simply an ecological theory subsequently expanded to accommodate social dynamics, was influential to new arts of government informed by systems and complexity sciences from its inception. The budworm even made an appearance in Howard Raiffa's lecture materials at Harvard Business School (Hutter 2018).

With regard to environmental governance, resilience has been part of a broader shift toward measuring and managing ecosystem functions rather than resource stocks, and in which the object of environmental management has become not so much individual species as social-ecological relations. This shift – exemplified by the now-ubiquitous management of natural systems as providers of 'ecosystem services' – is one in which environmental processes are instrumentalized as providers of critical functions sustaining (certain) human lives, and in which environmental governance becomes a key arena for the exercise of biopower. While the concepts of resilience and ecosystem services would not come into explicit contact until the 1990s, we can see in Holling's early work an emphasis on ecosystem function and an understanding of natural resources in terms of a 'capital inventory,' whose quantity and quality determine the resilience of the system (Holling 1974b: 60). By shifting the focus of ecosystem management from quantities of populations to qualitative relations among system elements, resilience promotes an environmental type of intervention that seeks to govern social-ecological systems in their relationality, while at the same time giving ecological management a central role in governmental practice.

While resilient strategies of management reflect Foucault's observations about American neoliberalism – namely its operation as an environmental type of biopower that governs populations by intervening in the conditions of life – its trajectories at IIASA also show that this mode of government was not specific to neoliberalism. As Rindzeviciute (2016) writes, systems analysis held ambivalent implications for the Soviet system in the 1980s and 1990s, smuggling in revolutionary elements even as it seemed to promise greater control. These techniques, she shows, were not simply imposed from the West but co-produced through 'the invention of 'common problems' requiring new modes of sharing data [...] and new institutional frameworks for action' (2016: 207). She argues that the post-socialist transition actually resulted in the decline of systems analysis techniques and the inversion of their revolutionary potentials; if systems analysis was implicated in neoliberal reforms, this was due more to institutional momentum than to any inherent 'fit' between these ideologies, as systems analysts were best situated within networks of power to oversee the transition. The association of decentralized control with neoliberal capitalism was not a foregone conclusion, but was the outcome of a deliberate strategy on the part of the international right that 'effectively decoupled the idea of the socialist system from the idea of the market, thus rendering the combination of these two ideas politically bankrupt' (Rindzeviciute 2016: 214; Bockman 2011).

IIASA itself was not insulated from neoliberal transitions. Holling's tenure as IIASA director in the early 1980s coincided with a hardening of international relations between the US and USSR, which precipitated a crisis at IIASA as US President Ronald Reagan withdrew funding from the institute, followed by the UK (Rindzeviciute 2016). The collaboration between IIASA and Bariloche on the 'surprise' project, which would have expanded the role of Third World researchers and research agendas in IIASA collaborations, was ultimately a casualty of 'IIASA overload' in this period (Holling 1984). In its place, new articulations of systems analysis and policy would take shape as the institute became a hub for the implementation of post-socialist market reforms in the early 1990s (Rindzeviciute 2016: 215).

This history suggests that if there is an empirical diversity of resiliences in the present (Anderson 2015), then this diversity was also present in the concept's origins. Reflecting on this diversity becomes all the more important insofar as resilience has shaped our understanding and experience of the Anthropocene as a generalized condition of environmental vulnerability, a global 'turbulence' to which we can only hope to adapt, through its articulation with neoliberal forms of market rule. Rather than revealing the resemblance between resilience and emerging neoliberal arts of government, this history might indicate their dissemblances, and the possibility of their undoing.

Notes

1 See Zebrowski (2014) for a discussion of resilience in the context of the evolving conceptions of nature in political economy and economics.

2 Major works on the crisis of the 1970s and post-Fordist transition include Harvey (2005), Magdoff and Sweezy (1972), and Arrighi (2010).

3 Perhaps the strongest articulation of this is the 'libertarian paternalism' of the behavioral economists Thaler and Sunstein (2008).

4 Holling's early work was squarely situated within debates on environmental crisis. In the 1970s an increase in budworm outbreaks, with disastrous effects on the Canadian forestry sector, became the subject of intense media attention and debate. At the time, the conventional management approach involved aerial spraying of DDT, a financially and ecologically costly tactic that became the impetus for Rachel Carson's *Silent Spring*, a flagship text for the modern environmental movement. The backlash against and subsequent ban of DDT, combined with the prohibitive cost of aerial spraying, prompted funding for research initiatives including the innovations in ecosystem modeling that would prompt Holling's invitation to join IIASA (Hutter 2018).

5 Peccei personally facilitated the intergovernmental talks that would result in IIASA's founding and served as the Italian representative on its member council (Pauli 1987; Bundy 1969a and 1969b).

6 Rindzeviciute (2016: 8) points out that 'when the term "governmentality" began to circulate in the early 1980s, the notion of systems analysis as an *art of governance* was being widely promoted in management education.' Thus she argues that Foucauldian governmentality studies themselves 'might be seen as a part of the system-cybernetic world of governance' (2016: 9).

7 For instance the Proposed Research Program for Management of Urban Systems, included in the conference proceedings, identified the goal of maximizing resilience as a central focus of urban design research to begin in 1974: 'the most important contribution of this work would be to help shift the mind-set of present-day economic and physical planners away from single-point 'optimal' or equilibrium-centered modes of thought' (IIASA 1973: 124).

8 The evolution of these approaches and how they have shaped applications of resilience in adaptive management deserves further investigation, in particular in the current use of 'decision theaters' in applications of resilience thinking where dynamic programming is employed alongside simulations. Thanks to Kevin Grove for this point.

9 See for instance the work of Derk Loorbach (2007) and Frank Geels (2010).

10 Here Holling echoes Spinoza's (in)famous formula 'God, or Nature.' Like Spinoza, for Holling it is the interconnection of all things – in his terms, their complexity – that is the source of nature's indeterminacy.

11 Schrickel (2014: 18) describes Häfele as one of the most controversial politicians in West Germany at that time.

12 These tensions are evident for instance in corporate forays into resilience, where reductionist ways of thinking and streamlined just-in-time supply chains run up against notions of complexity that prioritize redundancy, modularity of components, and long-term investments in resilience strategies whose returns may be nonlinear and unpredictable (Kupers, personal interview, March 15, 2016; Arizzi et al. 2014: 127).

13 In a 1975 visit to IIASA Jimmy Davidson, the Head of Group Planning at Royal Dutch Shell and early developer of the company's influential scenarios approach, was reportedly inspired by the work on resilience (Kupers 2014: 10).

References

Anderson, B. (2015) What kind of thing is resilience?' *Politics*, 35(1): 60–66.

Arizzi, S., Egger, M., Rittenhouse, D., and Williams, P. (2014) *Corporations and Resilience. In Turbulence: A Corporate Perspective on Collaborating for Resilience*. Amsterdam: Amsterdam University Press.

Arrighi, G. (2010) *The Long Twentieth Century: Money, Power, and the Origins of Our Times*. New York: Verso.

Arthur, B. (2014) *Complexity and the Economy*. Oxford: Oxford University Press.

Bell, D. E. (1977) A decision analysis of objectives for a forest pest problem. In D. E. Bell, R. L. Keeney, and H. Raiffa (eds), *Conflicting Objectives in Decisions*. New York: John Wiley & Sons, 389–424.

Bell, D. E., Keeney, R. L., and Raiffa, H. (eds) (1977) *Conflicting Objectives in Decisions*. New York: John Wiley & Sons.

Bockman, J. (2011) *Markets in the Name of Socialism: The Left-Wing Origins of Neoliberalism*. Redwood City, CA: Stanford University Press.

Bundy, M. (1969a) Memorandum of conversation with Dr. Dzermen Gvishiani, 3 April. Laxenburg, Austria: IIASA Archives.

Bundy, M. (1969b) Memorandum to: Dr. Henry Kissinger, 4 April.Laxenburg, Austria: IIASA Archives.

Carson, R. (1962) *The Silent Spring*. New York: Houghton Mifflin.

Chandler, D. (2014) *Resilience: The Governance of Complexity*. New York: Routledge.

Clark, W. (1975) Notes on certainty, uncertainty, and the unknown, 5 July. In W. Clark and H. Swain (eds), *Hypotheticality, Resilience and Option Foreclosure: Summary Notes of a IIASA Workshop* (IIASA Working Papers No. WP-75–080). Laxenburg, Austria: International Institute for Applied Systems Analysis, 27–30.

Clark, W. (2018) Telephone interview, 11 December.

Collier, S. (2011) *Post-Soviet Social: Neoliberalism, Social Modernity, Biopolitics*. Princeton, NJ: Princeton University Press

Ehrlich, P. (1968) *The Population Bomb*. New York: Ballantine.

Evans, B., and Reid, J. (2014) *Resilient Life: The Art of Living Dangerously*. Cambridge: Polity Press.

Fiering, M. B., and Holling, C. S. (1974) *Management and Standards for Perturbed Ecosystems* (IIASA Research Reports No. RR-74–73). Laxenburg, Austria: International Institute for Applied Systems Analysis.

Foucault, M. (2008) *The Birth of Biopolitics: Lectures at the College de France, 1978–79*. (G. Burchell, trans.). New York: Palgrave Macmillan.

Gallopín, G. (1980) Letter to C. S. Holling, 20 May. Laxenburg, Austria: IIASA Archives.

Gallopín, G. (1981) An institution in a turbulent environment: the Fundación Bariloche. Laxenburg, Austria: IIASA Archives.

Geels, F. (2010) Ontologies, socio-technical transitions (to sustainability), and the multi-level perspective. *Research Policy* 39(4): 495–510.

Grove, K. (2018) *Resilience*. New York: Routledge.

Grümm, H. R. (ed.) (1975) *Analysis and Computation of Equilibria and Regions of Stability, with Applications in Chemistry, Climatology, Ecology, and Economics. Record of a Workshop July 21– August 1*. Laxenburg, Austria: International Institute for Applied Systems Analysis.

Grümm, H. R. (1977) Talking about resilience. *Options Magazine*, Spring: 5–6. International Institute for Applied Systems Analysis.

Häfele, W. (1981) *Energy in a Finite World: A Global Systems Analysis*. Report by the Energy Systems Program Group of the IIASA. Cambridge: Ballinger.

Häfele, W., Balinski, and Holling, C. S. (1975) Correspondence: Resilience Workshop, 29 September – 26 November. Laxenburg, Austria: IIASA Archives.

Harvey, D. (2005) *A Brief History of Neoliberalism*. Oxford: Oxford University Press.

Holling, C. S. (1973a) Resilience and stability of ecological systems. *Annual Review of Ecology and Systematics* 4: 1–23.

Holling, C. S. (1973b) *Systems Resilience and Its Policy Consequences: Proposal for Work at IIASA* (IIASA Working Papers No. WP-73–71). Laxenburg, Austria: International Institute for Applied Systems Analysis.

Holling , C. S. (1973c) Chairman's recommendations and proposed research program. In IIASA, *Proceedings of IIASA Planning Conference on Ecological Systems, September 4–6, Vol. I: Summary and Recommendations* (IIASA Conference Proceedings No. PC-73–06). Laxenburg, Austria: International Institute for Applied Systems Analysis, 3–24.

Holling, C. S. (1973d) *On Leaving China* (IIASA Working Papers No. WP-73–76). Laxenburg, Austria: International Institute for Applied Systems Analysis.

Holling, C. S. (1973e) *On Making a Marriage, an Institute, or a Society* (IIASA Working Papers No. WP-73-l0). Laxenburg, Austria: International Institute for Applied Systems Analysis.

Holling, C. S. (1974a) *Notes Towards a Science of Ecological Management* (IIASA Working Papers No. WP-74–22). Laxenburg, Austria: International Institute for Applied Systems Analysis.

Holling, C. S. (1974b) *Ecology and Environment Status Report* (IIASA Status Reports No. SR-74–72-EC). Laxenburg, Austria: International Institute for Applied Systems Analysis.

Holling, C. S. (ed.) (1978) *Adaptive Environmental Assessment and Management*. New York: John Wiley & Sons.

Holling, C. S. (1984) Letter to John Steele, 3 February. Laxenburg, Austria: IIASA Archives.

Holling, C. (2001) Understanding the complexity of economic, ecological and social systems. *Ecosystems* 4: 390–405.

Holling, C. S., Häfele, W., and Walters, C. (1975) IIASA workshop on hypotheticality, resilience, and option foreclosure. Memorandum 27 June. In W. C. Clark and H. Swain (eds), *Hypotheticality, Resilience and Option Foreclosure: Summary Notes of a IIASA Workshop*. IIASA Working Paper WP-75–080. Laxenburg, Austria: International Institute for Applied Systems Analysis.

Hutter, M. (2018) Ecosystems research and policy planning: revisiting the Budworm Project (1972–1980) at the IIASA. In M. Christian, S. Kott, and M. Ondrej (eds), *Planning in Cold War Europe: Competition, Cooperation, Circulations (1950s–1970s)*. Berlin: Walter de Gruyter, 261–284.

International Institute for Applied Systems Analysis (IIASA) (1973) *Proceedings of IIASA Planning Conference on Ecological Systems, September 4–6, Vol. I: Summary and Recommendations*. (IIASA Conference Proceeding No. PC-73–6). Laxenburg, Austria: International Institute for Applied Systems Analysis.

Jones, D. D. (1977) Catastrophe theory applied to ecological systems. *Simulation* 29(1): 1–15.

Jones, D. D., Holling, C. S., and Peterman, R. M. (1975) *Fail-safe vs. Safe-fail Catastrophes* (IIASA Working Papers No. WP-75–93). Laxenburg, Austria: International Institute for Applied Systems Analysis.

Kupers, R. (ed.) (2014) *Turbulence: A Corporate Perspective on Collaborating for Resilience*. Amsterdam: Amsterdam University Press.

Loorbach, D. (2007) *Transition Management: New Mode of Governance for Sustainable Development*. Utrecht: International Books.

Lovelock, J. E., and Margulis, L. (1974) Atmospheric homeostasis by and for the biosphere: the Gaia hypothesis. *Tellus* 26: 2–10.

MacKinnon, D., and Derickson, K. D. (2012) From resilience to resourcefulness: a critique of resilience policy and activism. *Progress in Human Geography* 37(2): 253–270.

Magdoff, H., and Sweezy, P. M. (1972) *Economic History as It Happened, Vol. 1: The Dynamics of US Capitalism: Corporate Structure, Inflation, Credit, Gold, and the Dollar*. New York: Monthly Review Press, 197–212.

McDonald, A. (No date). Nuclear reactors and energy models [Weblog]. Retrieved from http://www.iiasa.ac.at/web/home/about/alumni/Alan_McDonald.html.

Meadows, D. H., Meadows, D. L., Randers, J., and BehrensIII, W. W. (1972) *The Limits to Growth: A Report for the Club of Rome's Project on the Predicament of Mankind*. New York: Universe Books.

Mirowski, P. (2002) *Machine Dreams: How Economics Became a Cyborg Science*. Cambridge: Cambridge University Press.

Mitchell, T. (1998) Fixing the economy. *Cultural Studies* 12(1): 82–101.

Nelson, S. (2014) Resilience and the neoliberal counter-revolution: from ecologies of control to production of the common. *Resilience: International Policies, Practices and Discourses* 2(1): 1–17.

Nelson, S. (2015) Beyond the limits to growth: ecology and the neoliberal counter-revolution. *Antipode* 47(2): 461–480.

Pauli, G. A. (1987) *Crusader for the Future: A Portrait of Aurelio Peccei, Founder of the Club of Rome*. Oxford: Pergamon Press.

Peccei, A. (1969) *The Chasm Ahead*. London: Collier-Macmillan Ltd.

Pelling, M. (2011) *Adaptation to Climate Change: From Resilience to Transformation*. New York: Routledge.

Raiffa, H. (2011) *Memoir: Analytical Roots of a Decision Scientist*. CreateSpace Independent Publishing Platform.

Rindzeviciute, E. (2016) *The Power of Systems: How Policy Sciences Opened Up the Cold War World*. Ithaca, NY: Cornell University Press.

Schmidt, J. (2015) Intuitively neoliberal? Towards a critical understanding of resilience governance. *European Journal of International Relations* 21(2): 402–426.

Schrickel, I. (2014) Von Schmetterlingen und Atomreaktoren: Medien und Politiken der Resilienz am IIASA. *Behemoth* 7(2): 5–25.

Swain, H. (1975) Rapporteur's notes for workshop on resilience, hypotheticality, and option foreclosure. In W. Clark and H. Swain (eds), *Hypotheticality, Resilience and Option Foreclosure: Summary Notes of a IIASA Workshop* (IIASA Working Papers No. WP-75–080). Laxenburg, Austria: International Institute for Applied Systems Analysis, 5–14.

Thaler, R., and Sunstein, C. (2008) *Nudge: Improving Decisions about Health, Wealth, and Happiness*. New Haven: Yale University Press.

van der Meyden, H. (2014) Nexus! Resilience in a pressure cooker. In R. Kupers (ed.), *Turbulence: A Corporate Perspective on Collaborating for Resilience*. Amsterdam: Amsterdam University Press, 87–100.

Walker, J., and Cooper, M. (2011) Genealogies of resilience: from systems ecology to the political economy of crisis adaptation. *Security Dialogue* 42(2): 143–160.

Zebrowski, C. (2014) The nature of resilience. *Resilience: International Policies, Practices and Discourses* 1(3): 159–173.

7

DESTITUTING RESILIENCE

Contextualizing and contesting science for the Anthropocene

Kevin Grove and Allain Barnett

Introduction

Few concepts have proven to be as politically paradoxical for critical social scientists as *resilience* and the *Anthropocene*. For the former, resilience holds out the promise of novel ways of conceptualizing human-environment relations, and managing nature-society interactions, which are founded on a holistic and relational ontology of complexity, interconnection and emergent becoming rather than high modernist separation and division. But critical scholars have also argued that resilience thinking often supports a conservative and reactionary environmental politics that shores up the ongoing neoliberalization of socio-ecological relations. For the latter, the Anthropocene's imaginary of a new geological epoch driven by human impacts on Earth systems radically destabilizes the Kantian split between consciousness and Earth that has stabilized the modern subject for nearly three centuries (Grove and Chandler 2017). But this sublime vision that couples human activities to planetary dynamics is also inspiring new technopolitical geo-engineering fantasies that reinforce a modernist vision of the world as an object of totalizing human knowledge and predictive control (Neyrat 2019)

This chapter explores how the ontologies of complexity, emergence and interconnection that underpin visions of resilience and the Anthropocene can paradoxically reinforce a modernist split between nature and society, recalibrate modernist fantasies of total control and yet still open onto novel political possibilities. We focus here on the way advocates of each concept reassert ontological foundations that attempt to ground historically specific understandings of nature, society, politics and ethics in a renaturalized vision of reality. Whether this is resilience proponents' visions of hierarchically organized, institutionally mediated complex social and ecological systems (Holling 2001) or Anthropocene imaginaries of hybrid, interconnected socio-ecological assemblages (Latour 2004), these ontological affirmations attempt to carve out a new level of reality that is beyond critical contextualization. Our goal here is to build on recent critical engagements that have begun to historicize these ontological claims. Zeroing in on ecological resilience thinking, we detail how ecologists' visions of complexity have their roots in the mid-twentieth-century work of cybernetic behavioural scientist Herbert Simon (Grove 2018). As we will detail, Simon's Nobel Prize-winning work on what he called 'bounded rationality' recalibrated liberal understandings of rationality for a complex world that exceeded the sovereign individual's cognitive capacities. Simon transformed liberalism's calculatory 'will to truth' – the drive and desire to apprehend 'what exists' as objectively knowable entities amenable to predictive control – into a *will to design*, the drive and desire to apprehend 'what exists' as rational abstractions amenable to pragmatic, functional synthesis (Grove 2018). While resilience expresses, in Agambenian terms, the maximum *application* of a de-Naturing cybernetic will to design, its deployment in contextually specific struggles against racialized forms of violence and suffering that sustain the Anthropocene often renders the term *inoperative*, available for creative use for any number of political projects that are irreducible to the cybernetic management of complexity.

This chapter begins by excavating the ontological claims that support contemporary thinking on the Anthropocene and resilience. After demonstrating the tendency for processual ontologies of hybridity, emergence and socio-ecological becoming to open onto what Neyrat (2019) describes as new fantasies of totalizing geo- and eco-constructivist control, we detail how these fantasies entered ecological research through resilience ecologists' engagements with Herbert Simon's thought on both the hierarchical organization of complexity, and its implications for rational decision-making. The chapter concludes with an extended discussion of the possibilities for destituting resilience and contesting the de-Naturing accounts of resilience and the Anthropocene that have captured both applied and critical scholars' imaginations in recent years.

(Against) Anthropocene ontologies

Although the term 'Anthropocene' developed in the Earth sciences, its provocation to envision humans as geological force has intensified critical efforts to rethink human-environment relations across the social sciences. The concept strikes an apparent deathblow against the Cartesian distinctions between body/mind, world/consciousness, and nature/humanity that have structured modernist thought over the past four hundred years (although we will see below that this is not necessarily as clear-cut as might seem). An Anthropocene imaginary does not position (modern, Western) humanity in a privileged, 'God's-eye' position outside and above the world. To be sure, some versions envision the Earth as a techno-ecological object, encrusted with interwoven infrastructure systems that stretch from mines extending deep in the planetary crust, to communication satellites orbiting far beyond the stratosphere (see Amin and Thrift 2017). But alongside these ongoing technologically mediated interventions into Earth systems, the Anthropocene also foretells of a future

Earth that has become uninhabitable and devoid of human life (Evans and Reid 2014). Such visions of an Earth that is indifferent to human life – and thus indifferent to the existence of thought itself – challenges scholars to rethink key categories of ethical and political thought in ways that convey more humble images of humanity. Rather than the passive backdrop on which modern political drama plays out, the Anthropocene reveals that the Earth has always actually been an agent, conditioning human possibilities on timescales that exceed a modernist imaginary (Latour 2017; Tsing 2015).

What David Chandler (2018) calls 'affirmations of the Anthropocene' give 'life itself' ontological value: Anthropocene visions of a dynamic Earth translate easily into ontologically charged visions of lively materiality that precedes human consciousness and wilful agency. This is particularly clear in the emergence of new materialist thought. For example, Jane Bennett's (2010) influential slant on 'vibrant matter' situates agency across material-discursive assemblages that condition how material and ideational effects actualize out of ontologically prior relational capacities. Relatedly, Elizabeth Grosz's (2011) feminist metaphysics reconceptualizes nature, culture, ethics and politics in relation to an irreducible and ontologically prior materiality that is not simply an unthinkable excess but rather a vital force that shapes potentiality. Such arguments position a vibrant, vital materiality as an ontological foundation that determines the possibilities for both ethics and politics. This renaturalization of politics attempts to open thought to new possibilities attendant on affirming rather than immunologically negating the emergent connectivity and relationality that precedes the sovereign subject. Critical work on the Anthropocene has taken up these possibilities through examining how the concept might provoke new understandings of politics and the political. For example, Yusoff (2013) offers an image of geo-politics that positions the Earth – 'geo' – as the condition for political life. Rather than the ontological foundation of political life, the agential subject here becomes one expression of planetary potentiality. And rather than the stable backdrop of human politics, earthly dynamics shape the possibilities for politics (see also Clark and Yusoff 2017). The affirmation of the Anthropocene thus challenges scholars to develop new concepts and categories for understanding how human action emerges *through* and *with* – rather than on and against – a revitalized Earth.

An Anthropocene imaginary thus opens provocative epistemological and ethical questions. First, how might critical scholars pry open the concepts and categories of social thought – which have long assumed the Earth to be a stable backdrop for human affairs (Dalby 2013) – to forces, magnitudes and durations that comprise Earth's movements? As Nigel Clark and Kathryn Yusoff (2017) emphasize, these modes of thought have long concerned the geological sciences, which have recognized for decades the planet's inherent dynamism. But key categories of social thought that have structured modern liberal societies, such as agency, sovereignty, subjectivity, nature and power, have yet to be recalibrated for this new reality. Similarly, in ethico-political terms, how to cultivate respect for *asymmetrical* relations with a vital Earth that conditions the possibilities for human life itself? For

Clark (2010), this involves in part recognizing humanity's geo-historical indebtedness to previous modes of living, and particularly Aboriginal and Indigenous peoples, who cultivated habitable environments out of an otherwise indifferent planet.

At stake in recent reassertions of ontology is how visions of lively and dynamic materiality might figure into ongoing progressive, radical and anti-essentialist efforts to reimagine collective possibilities in the wake of the failures of 1968 and the wider destabilizing of the Marxist ontology that envisioned a unified proletariat as the subject of revolution (Chandler 2014; Barnett 2017). There is much to be appreciated in this ontological turn, especially the possibilities it holds out for new imaginaries of power that allow us to reconceptualize how liberal governance structures its relation with incommensurable difference (Povinelli 2011, 2016) and fluctuates through material and infrastructural volumes (Steinberg and Peters 2015). However, as Chandler's (2018) exploration of what he calls 'ontopolitics in the Anthropocene' demonstrates, new modes of governance are *also* emerging on the back of these ontological claims.

The term *ontopolitics* foregrounds the often underspecified ontological assumptions that structure governmental rationalities. For example, centralized command-and-control management strategies that have typified modernist planning rely on ontological assumptions that 'nature' is more or less stable, linear and predictable; that humans occupy a 'god's-eye-view' external to natural dynamics; and that human rationality can identify transcendent laws that determine social and environmental outcomes and bend these external realities to individual will. Of course, Anthropocene imaginaries call into doubt this modernist ontology of separation, sovereignty and objective rationality. But Chandler's point is that the emergence of new modes of governance organized around seemingly progressive or radical strategies such as 'resilience' or 'hacking' contain their own under-acknowledged ontological commitments that blind scholars and practitioners to their partiality and situatedness. The danger here is that, while critical scholars' affirmation of the Anthropocene might *promise* a radical break with both essentializing modernist ontologies *and* poststructural suspicions of ontological claims, it also continues a Western tradition of devaluing the social world of everyday practices, strategies and struggles that complicate the elegant lines of ontological reasoning. As Barnett (2017) cautions, any effort to ground understandings of ethics and politics on rarefied forms of philosophical reasoning risks blinding critical analysis to the mundane, practical situations in which people take up concepts such as resilience or the Anthropocene in a variety of disconnected and contradictory ways in response to contextually specific problems. The stakes are significant: while on an ontological level the Anthropocene appears to introduce discontinuous breaks with prior modes of thought and practices of control; an analytics of problematic situations highlights subtle continuities and topological reconfigurations that extend and intensify prevailing power relations for novel socio-ecological contexts (Collier 2009).

A growing body of critical scholarship is beginning to historicize these ontological affirmations. For example, Lynne Huffer (2015) cautions against the recent turn in feminist thought, outlined above, to vitalist-infused materialism. In her reading, the concern with 'life itself' is a problem unique to the Anthropocene age that needs to be unpacked. Posing the question of how 'life itself' became a problem concerning leading feminist thinkers (rather than positioning 'life itself' as an irreducible reality on which a theory of ethics and politics can be subsequently articulated) directs analytical attention to shifting configurations of forces, techniques and practices that structure contextually specific understandings of life. In her reading, the Anthropocene does *not* force humans to confront an ontologically prior relationality and earthly vitalism; rather, it confronts Western subjects with a material archive – the fossil record – that destabilizes modern subjectivity. For Huffer, the fossil, like unreason, is a 'privileged locus of a resemblance out of sync with the time of its appearance', a heterotopic space that catalogues a violent history of catastrophic extinctions that subtends any apparent 'irreducible' vitality. From this perspective, the Anthropocene does not provoke a life-affirming vitality, but rather returns thought to the founding split between subject and world that inaugurates modern, Western subjectivity. The affirmation of the Anthropocene simply inverts Kant's epoch-defining turn away from the sublime: if Kant, reflecting on the 1755 Lisbon earthquake, constructed a vision of consciousness separated from world to shelter the human from unthinkable disaster, affirming the Anthropocene immerses the human *within* this catastrophe-prone world. And rather than elaborating a theory of ethics and politics founded on the modern sovereign individual, the turn to new materialist and vitalist thought attempts to elaborate a theory of ethics and politics founded on the always-already embedded, 'worldly' individual. What is lost in this move parallels what Clark (2010) identifies as the asymmetrical relation between Earth and humans – but where Clark emphasizes human indebtedness to sublime earthly forces, Huffer and other critics focus on a second asymmetry, what Colebrook (2012) describes as humans' *sym-thanotic* evolution – or the way human evolution has occurred through, rather than in spite of, systematic destruction. Reading across Heidegger and Derrida, Colebrook (2012) suggests the Anthropocene emerges out of a defining feature of modern subjectivity: the uniquely human capacity to bracket existence and think beyond the given. Of course, this ability to imagine the world as such conditions the technical character of human being, for it offers a vision of reality comprised of objects given for instrumental human use – including 'the world' itself. But it *also* conditions the *ecological* character of human being, for it offers an image of world as 'environment', a milieu or dwelling (or *oikos*) given to life itself.

In this light, the Anthropocene signals an 'irrevocable and inhuman humanity': a humanity that has detached itself from its animality and lives life in a way that is destructive of its milieu (Colebrook 2012). The Anthropocene does not express a violent Earth that destabilizes the modernist ontological division as much as an intensification of the destructive tendencies of the modern subject. The modern human appears here less a rational animal than a geological event: not one being amongst others immersed in an interconnected world, but rather a life-annihilating vector. Of course, the destructive character of European modernity has long been a major theme for critical thought. Adorno and Horkheimer's *Dialectic of Enlightenment*, Benjamin's

'On the Concept of History', and Foucault's (2003) inversion of Clauswitz's formulation of war as politics by other means all recognize, in their own way, how the avowed 'universal peace' of liberal rule rests on subjugated histories of death and destruction of non-European lives and forms of life. Critical race theory has similarly shown how the supposed universal values of humanism require the production of less-than-human lives, black and indigenous lives devoid of value and significance (e.g., Sexton 2011). Indeed, the modern experience of anticipatory temporality – the future as terrain of limitless growth and development – was made historically possible through the racialized violence of the plantation economy that *de*-personalized and instrumentalized bodies of African origin marked as black (McKittrick 2013; Thomas 2016).

Frederic Neyrat (2019) finds this destructive tendency repeating itself in dominant strains of what he calls *constructivist* Anthropocene fantasies. His sense of constructivism does not reference decades-old debates over the social construction of reality; rather, he emphasizes how otherwise disparate responses to the Anthropocene attempt to render Earth an object of technocratic management. This is perhaps most apparent in what he calls *geo-constructivist* visions of a 'good Anthropocene' found in geo-engineering, climate engineering and ecological modernization proponents' assertions that human technological ingenuity will master dynamic and uncertain Earth system processes, thus saving the Earth from its own destructive tendencies. But it is also present in what he calls *eco-constructivist* promises of ecologists and environmental scientists to adaptively manage complex ecological, biological and geochemical system dynamics in ways that will sustain valued ecosystem services, whatever these may be. Efforts to artificially engineer planetary systems and adaptively manage ecosystems share a common drive to detach humans from their organic milieu: both eco-constructivist visions of complex ecological networks and geo-constructivist visions of planetary-wide manufactured living environments hinge on apprehending the environment as nothing more than an assemblage of abstract objects whose essence lies in their functionality. For example, decades of work in ecology have shown that, contrary to popular environmentalist rhetoric that 'everything is connected to everything,' ecosystems are structured by a handful of key functional relations between systemic elements. The specific identity of each element matters less, from an ecological point of view, than the functional role it might play within a wider system (Grove 2018).

This creation of a 'third nature' – that is, 'nature' as functional environment amenable to human design – extends to the level of consciousness. In contemporary infrastructurally mediated worlds of liberal life, subjective experience – and memory itself – is increasingly externalized as data (Stiegler 2018). This 'neuro-liberalism' involves the ubiquitous presence of smart devices that facilitate an increasingly broad range of everyday activities, transforming a near-infinite number of interactions with surrounding environments into discrete data points that provide a new surface for governmental regulation (Jones et al. 2013). While the ongoing data revolution may allow for new forms of participatory and integrative governance initiatives, such as community-oriented resilience hubs or the resilience hackathons Chandler (2016; 2018) details, it furthers modernity's drive to construct a worldless humanity – in this case, a form of humanity whose worldly existence has been replaced by data streams amenable to modulation and reflexive cybernetic management (Neyrat 2019).

As Agamben (2016) cautions, at this point, language has ceased to function as a medium for the communication of ontology and has instead saturated being itself: the world of the Anthropocene, a world of and for humanity severed from any organic connection, is a world sensed and apprehended as essentially informational vectors. Of course, this is a tendency that Jean Baudrillard presciently cautioned against two decades before Earth scientists' naming of 'the Anthropocene.' Writing in a provocative chapter entitled 'Design and Environment,' Baudrillard (1981) connects an environmental aesthetic with the 1919 inception of the Bauhaus movement. While he is careful to situate the Bauhaus in a wider context of Europe's early twentieth-century technological disillusionment and cultural upheaval, he suggests that the Bauhaus articulated an experience of world as 'a rational conception of environmental totality' (1981: 186–7), or a functional system of objects amenable to human intervention and manipulation (Grove 2018). This experience involves a 'dissociation of every complex subject-object relationship into simple, analytic, rational elements that can be recombined in functional ensembles and which take on the status of the environment' (Grove 2018). Its implications cannot be overstated: the Bauhaus offers a vision of interiority and exteriority – subject and world, society and nature, security and anarchy – as objective entities amenable to functional synthesis. After the Bauhaus, the modern subject no longer grounds objective truth, but rather straddles the threshold between (subjective) self-image and (objective) self-as-image, an entity whose meaning and value is relationally determined and reflexively mutable. And nature no longer stands apart from this subject, an external and alien world at once amenable to rational, predictive control and the locus of sublime destructive forces; rather, as a functional 'environment,' nature becomes artifactual, a domain of functional essences whose complex interrelations escapes totalizing knowledge and yet remains open to designerly synthesis.

The Bauhaus thus consolidates a tendency, across a variety of contemporaneous social, political and aesthetic trajectories, to intensify the modernist division between humanity and nature. In de-mystifying art and developing a functional system of value (Foster 2002), the Bauhaus similarly de-mystified any sense of humanity's organic being. As Baudrillard (1981: 201) emphasizes,

> If one speaks of environment, it is because it has already ceased to exist. To speak of ecology is to attest to the death and total abstraction of nature … the great signified, the great referent Nature, is dead, replaced by environment, which simultaneously designates and designs its death and the restoration of nature as simulation model (its 'reconstitution,' as one says of orange juice that has been dehydrated).

The world of the Bauhaus is a world of objects – people, non-human organisms, material objects, technologies and so forth – apprehended as potential functions.

For Baudrillard (1981: 186–7), 'the whole environment becomes a signifier, objectified as an element of signification. Functionalized and liberated from all traditional implications (religious, magical, symbolic), it becomes the object of a rational calculus of signification.' This 'rational calculus of signification' introduces an *in-formational* twist to a modernist ontology, in a double sense of the term. First, because the essence of entities lies in their functionality, they are necessarily indeterminate, always-already in-formation. Second, what determines any given entity is the quality of information flow that structures relationships between entities and determines their functionality. On this basis, Baudrillard draws a provocative link between the designerly thought and practice of the Bauhaus movement, focused on the creation of functional artifacts stripped of ornamentation, and the emergence of cybernetics, the science of communication and control: both share a common *in-formational* understanding of complexity, a common apprehension of the world as a system of functional relationships amenable to synthesis and pragmatic control.

For example, Herbert Simon, a Nobel Prize-winning behavioural scientist whose work on decision-making in complex environments had a formative influence on early resilience ecologists (more on this below; see also Grove 2018), offers a cybernetic understanding of environment that strongly resonates with Baudrillard's characterization of the Bauhaus. Writing in an influential 1956 paper, Simon (1956: 130) argued that:

> The term *environment* is ambiguous. We are not interested in describing some physically objective world in its totality, but only those aspects of the totality that have relevance as the 'life space' of the organism considered. Hence, what we call the 'environment' will depend upon the needs, drives, or 'goals' of the organism, and upon its perceptual apparatus.

In Baudrillard's terms, Simon's characterization of environment as an emergent and contingent effect of the 'needs, drives or "goals"' of an organism reconstitutes Nature as a near-infinite series of functional relations whose complexity exceeds total human comprehension. But while this functional environment may elude totalizing objective knowledge and predictive control, the work of Simon and other cyberneticians (especially Norbert Weiner) developed a series of techniques for reflexively knowing and adaptively managing through complexity. These early forms of modulation refashion a liberal will to truth: the drive to reduce the world to objective truth amenable to predictive control becomes, in the hands of these cyberneticians, a drive to intervene in and adapt to a complex world from a position of necessarily partial or, to play on Simon's (1955) phrase, *bounded* knowledge.

We will explore the links between the Bauhaus, Simon and resilience ecologists below. For now though, Baudrillard's arguments here on the designerly 'death of Nature' and is 'reconstitution' in ecological simulation models helps further specify the historical context of Anthropocene thinking. Critical resilience scholars have increasingly recognized the influence of cybernetic thought on early resilience scholars (Walker and Cooper 2011; Chandler 2013; Zebrowski 2016; Grove 2018; Neyrat 2019; see also Nelson, this volume). Cybernetics, as the science of communication and control, offered ecologists a vision of ecology as complex systems mediated through flows of information – whether this is carbon, organic matter, 'indicator' species signalling ecosystem change or collapse, species movement, or some other medium of communication that facilitates feedback and reflexive control. Baudrillard's focus on the designerly roots of cybernetics allows us to link these insights to debates over the Anthropocene, for it compels us to explore how a designerly ethos inflects thought on both resilience and the Anthropocene. The next section thus unpacks the designerly roots of resilience ecologists' understanding of complexity.

The designerly roots of resilience

Like many ontologies of complexity, resilience ecologists tend to offer a vision of a single, unitary world characterized by complex, nonlinear interconnections between systemic components and uncertain, emergent change (see Fagan, this volume; Wakefield, this volume). However, if this ontology grasps 'the' world as unavoidably complex, this does not mean that complexity is undifferentiated across space. Rather, the complex world of resilience ecologists is a hierarchical system of systems, what C. S. Holling and colleagues term 'panarchy' (Holling 2001; Gunderson and Holling 2002). *Hierarchy* here does not refer to relations of control or authority. Instead, in Herbert Simon's influential formulation, hierarchy signals 'a system that is composed of inter-related sub-systems, each of the latter being, in turn, hierarchic in structure until we reach some lowest level of elementary subsystem' (Simon 1962: 468). Hierarchy is thus a way to understand organized complexity in systems in which there is no subordinate relation amongst sub-systems.

Ecologists directly borrowed Simon's hierarchical understanding of complexity to develop their theories of panarchy and adaptation. Holling and Sanderson (1996: 77–8) explain Simon's influence in an early discussion of panarchy:

> [Herbert] Simon was one of the first to describe the adaptive significance of hierarchical structures. He called them 'hierarchies,' but not in the sense of top-down sequence of authoritative control. Rather, semiautonomous levels are formed from the interactions among a set of variables that share similar speeds (and, we would add, geometries). Each level communicates a small set of information or quantity of material to the next higher (slower and coarser, in geometric terms) level. This 'loose coupling' between levels allows a wide latitude of 'experimentation' within levels, thereby increasing the speed of evolution.

This passage is repeated, almost verbatim, in several other essays written during the mid-1990s and early 2000s (see Grove 2018). Its influence on resilience ecologists' understanding of resilience cannot be overstated. In his field-defining formulation of adaptive management (see Nelson, this volume), Holling explains that,

> Simon has shown that such [hierarchical] structures have remarkable survival properties. First, removal of one sub-assembly does not necessarily destroy the whole. Because of the minimal connection between subassemblies, the others can persist, often long enough for self-recovery. Second, for the same reason, these structures rapidly adapt to change. As long as the same connections are maintained to other subassemblies, major changes and substitutions can take place within the sub-assembly. Species can substitute for species as long as the same function or role is performed.
>
> *(Holling 1978: 28)*

In effect, Simon's vision of complexity allows resilience ecologists to visualize species as systemic components of ecosystems that have been reduced to abstractions that can functionally substitute for one another. This is a significant transformation in scientific thought. What matters for resilience thinking is *not* the precise identity of individual elements within the system. This is the concern of an analytical natural science. Just as, through the Bauhaus, the essence of any entity lies in its functional relations with other entities, so too do resilience ecologists become concerned with the functional roles of species. In a resilient system, multiple species can serve the same functional role. A resilient ecosystem is thus characterized by biodiversity, for diversity allows other species to fill in and perform the same functions if an external shock or stress impacts one or more systemic components. But diversity has these effects only because species are nothing more than functional abstractions that are potentially substitutable for one another – a functionality expressed through ecological concepts such as *redundancy, keystone species, niches, connectivity*, and so forth.

An aesthetic of environmental totality thus lies at the heart of resilience ecology. This vision of environment as a complex, hierarchically organized functional system allowed ecologists to transform both environmental science and environmental management around the synthetic imperatives of designerly thought. Because complexity exceeds individual knowledge, science now uses techniques such as models to tease out functional relations and manage feedback loops in order to improve system performance (Grove 2018). As Holling explains while describing what this new ecological science should look like,

> the premise of this second stream [a science of the integration of parts] is that knowledge of the system we deal with is always incomplete. Surprise is inevitable. Not only is the science incomplete, but the system itself is a moving target, evolving because of the impact of management and the progressive expansion of the scale of human influences on the planet.
>
> *(Holling 1995: 13)*

In a world of complexity, there is no position from which scientists and environmental managers can attain totalizing, objective views of the system. Knowledge of the system is thus always limited (in Simonian terms, bounded) and emerges through the step-wise, adaptive process of interacting with the system: through designing interventions, monitoring how those interventions affect system performance and adjusting interventions based on the new information. But this is not the objective knowledge of modernist science. Rather, it is pragmatic knowledge that emerges through experience – experience that might be artificially crafted through simulation models (Holling 1978) or the real-world experience of active adaptive management programs (Lee 1993).

For resilience ecologists, complexity erases any qualitative or temporal distinction between knowledge-generating activities and environmental management. Because management involves an intervention that will disturb the system and generate previously unknown effects, any management activity will necessarily generate new knowledge about the nature of functional relations between component parts that structures a system. At first glance then, resilience ecology might offer a significant challenge to the modernist relation between science and management. If modernist thought offers a vision of science as the domain of objective truth and predictive knowledge, and management as the rational application of predictive knowledge to real-world problems, then a hierarchically organized complex world would seem to undermine this top-down management style. And indeed, resilience ecologists created a space for their visions of complexity and adaptive management through a devastating critique of command-and-control environmental management strategies (Walker and Cooper 2011; Grove 2018). But even as they kill off modernist fantasies of predictive control, resilience ecologists reconstitute the promise of control in cybernetic terms. And once again, Simon's vision of decision-making in complex conditions provided their blueprint. Simon's formulation of hierarchy emerged out of his long-running interest in empirically understanding decision-making processes.

Just as resilience scholars would later critique top-down understandings of environmental management that assumed a universally valid understanding of 'rationality,' so too did Simon's work develop out of his sympathetic critique of universalizing visions of rationality found in neo-classical economics and first-order cybernetics. Rather than assuming that human rationality followed an undifferentiated logical procedure, Simon (1955, 1956) argued instead that rationality was 'bounded' by the complex environments decision-makers were immersed within. These environments exceed the requirements for totalizing objective comprehension that structure universalizing visions of rationality: decision-makers cannot access all information about complex environments - violating neo-classical assumptions of perfect information. They also cannot process all accessible data and information, which means they cannot identify all functional relations in a way that enables prediction - violating the neo-classical equivalence between rationalization and optimization (Grove 2018). Individuals thus do not optimize, but rather pursue a course of action they expect to produce acceptable (if not optimal) results. Simon calls this *satisficing* behaviour – a boundedly rational and pragmatic behaviour that allows individuals to navigate complex environments that are inherently unknowable with any predictive precision.

Simon's boundedly rational individual is thus not the classically liberal figure of *homo economicus,* the rational decision-maker who chooses the optimal course of action based on perfect, predictive knowledge. Rather, the boundedly rational individual is an *adaptive* individual, *homo adaptivus* (Crowther-Heyck 2005). Adaptation is essential to Simon's sense of bounded rationality. Indeed, for Simon, to be adaptive is to be rational in a complex world. He equated the two terms, writing with his frequent collaborator, physicist Allen Newell, that:

> the behaviors commonly elicited when people (or animals) are placed in problem solving situations (and are motivated toward a goal) are called *adaptive*, or *rational*. These terms denote that the behavior is appropriate to the goal in the light of the problem environment; it is the behavior demanded by the situation.
>
> *(Newell and Simon 1972: 53)*

Critical social scientists have long recognized the functionalist roots of adaptation in fields such as human ecology and systems anthropology (e.g., Watts 2015). But Newell and Simon's formulation foregrounds how this vision of adaptation is linked to visions of a de-Natured, environmental totality with its roots in the Bauhaus, *and* bound up in recalibrations of liberal rationality for a world of hierarchically organized complexity. They go on to suggest that,

> if there is such a thing as behavior demanded by a situation, and if a subject exhibits it, then his [sic] behavior tells more about the task environment than about him ... if we put him in a different situation, he would behave differently.
>
> *(Newell and Simon 1972: 53)*

Homo adaptivus is thus a topological figure that (in Newell and Simon's formulation) straddles the threshold between an external environment that poses problems, and an internal consciousness that processes information from this environment and responds in an appropriate manner (Grove 2018). It does not stand apart from the world with the Cartesian individual's 'god's-eye view', but rather is always-already embedded in a complex world that presses on the subject (Chandler 2013). The adaptive subject is always at once inside and outside, never fully removed from its environment yet always reflecting on and adapting to the problems this environment poses. The process of decision-making is precisely the 'site' where she engages with a complex environment and adjusts her decisions – adapts – in order to realize a goal.

Simon's vision of bounded rationality and adaptation is thus less a decisive *break* with the modern individual than a *reformulation* of liberal rationality for a complex world. Adaptation spatially and temporally re-distributes rationality: in terms of the former, an adaptive rationality is not centred on an individual will, but is rather dispersed across the threshold between an individual and its external complex environment; in terms of the latter, rationality is not concentrated in the moment of decision that expresses the individual's sovereign will, but rather is distributed across a reflexive decision-making process. Similarly, ecologists' formulation of adaptive management does not offer a decisive break with modernist fantasies of totalizing control as much as it recalibrates how this control might be achieved in a complex world. This is a world that is no longer apprehended through an onto-logical split between subject and object (of knowledge, of control), but rather a world apprehended as hierarchically organized complexity: a world of functional essences given to pragmatic, adaptive human use. But this use does not result from individual interests; rather, it reflexively emerges as individuals reflect on and adapt to new and unexpected problems complex environments pose to their ability to realize their goals, whatever these may be.

This brief history of resilience ecologists' adoption and adaptation of Herbert Simon's thought on complexity and rationality thus provides an additional fold to debates on the Anthropocene detailed above. While the forms of knowledge pro-duction and management strategies that have emerged through ontologies of complexity have the effect of intensifying rather than destabilizing the modernist drive to detach humanity from the non-human, this intensification has been facilitated in part through the way scientists and practitioners in a number of fields – but especially ecology – have critically recalibrated key tenets of modern science and liberal political philosophy in relation to the functional, environmental totality the Bauhaus consolidated (Grove 2018). Geo-constructivist attempts to engineer resilient planetary systems and eco-constructivist attempts to design and manage resilient ecologies (Neyrat 2019) are both grounded, as it were, in the death of Nature that Baudrillard locates in the Bauhaus' functional rendering of world as 'environment:' both emerge out of the cybernetic reconstitution of Nature as a functional environment amenable to adaptive regulation and control. And as Neyrat (2019) and others caution, from this vantage point, the politics of resilience appear particularly bleak, as any attempt to build a political program for the Anthropocene upon ontologies of interconnection and complexity risks intensifying rather than escaping a modernist drive to totalizing control – whether this is a liberal will to truth or cybernetic will to design. However, as the final section details, it is arguably at this point where the promise of modernity reaches its cybernetic apogee that it becomes inoperative, a site of contestation, tension and conflict rather than a smooth, seamless modulation of complex social and ecologi-cal system dynamics.

Inoperative resilience

As resilience has become an increasingly influential organizing concept in a number of disparate fields over the last decade, critical scholarship on the concept has become increasingly dismissive. While early critical engagements with resilience thinking historicized its more brash universalizing claims and revealed ecologists' account of complexity to be a partial and situated response to wider social and environmental dynamics (Walker and Cooper 2011; Chandler 2014), there has also been a growing critical tendency to equate resilience thinking with neoliberal ideology. This equation relies on identifying formal similarities, such as a shared suspicion of centralized governance, a common valuation of insecurity, vulnerability and crisis as beneficial, and a shared emphasis on the importance of living with and profiting through risk (e.g., Watts 2015). The result has been a wave of critical research that sees resilience as an irretrievably compromised concept (MacKinnon and Derickson 2012).

And yet, despite these formal affinities, in practice resilience has been taken up in particular contexts by social and climate justice advocates in a number of surprising ways. Two examples will provide illustration. First, as Stephen Collier and colleagues demonstrate, recent urban resilience initiatives in New York City's Lower East Side have surprisingly earned the cautious support of local tenants' rights organizations that have, since the 1950s, been at the front lines of battles against gentrification and residential displacement (Collier et al. 2017). These seasoned activists' support of resilience thus cannot be credibly attributed to some kind of ideological duping that requires rarefied critical analysis to peel back the layers of obfuscations. Rather, these activists responded to what they saw as a strategic opportunity offered through the city's resilience initiatives in response to 2012's Superstorm Sandy. Specifically, Rebuild by Design – a US$950 million design-driven reconstruction project funded through the US government's Office of Housing and Urban Development and the Rockefeller foundation – deployed design techniques to incorporate community members in all stages of the project planning process, from conception to the ongoing (at the time of writing) implementation phase. Local activists felt these initiatives offered an opportunity for genuine public participation in community planning that had long excluded minority and impoverished residents from any decision-making capacity. But they cautioned that their support would turn into fierce opposition if they felt the city government was beginning to exclude them again (Collier et al. 2017).

Second, a similar dynamic animates urban resilience planning in Miami, where social justice and climate justice activists have seized on local governments' newfound concern with building the region's resilience to climate change impacts to gain unprecedented levels of access to and influence over public service provisioning. Miami's unique history of racially segregated development and racially and economically exclusionary governance has traditionally offered few avenues for public engagement, and has effectively dampened wider participation in urban governance for much of the past 60 years (Rose 2015). But local governments' recent participation in the Rockefeller Foundation's 100 Resilient Cities program and the City of Miami government's successful 2017 passage of a US $400 million bond offer has created opportunities for activists to place longstanding issues such as affordable housing and transportation at the forefront of the city's resilience planning agenda. Thus, while resilience planning undoubtedly recalibrates a variety of

techniques that have sustained segregated development and exclusionary governance, it also creates a narrow space where activists are able to make claims on local
authorities that, for the time being, are being recognized as legitimate issues worthy
of concern (Grove et al. 2018).

In both cases, the gains activists have made through seizing on the uncharted
territory of resilience planning – in the words of Henk Ovink, Rebuild by Design
principle, 'building the plane while flying it' (Cohen 2016) – are tenuous at best.
As the construction of the so-called Dryline in the Lower East Side winds its way
through the New York City government's permitting process, many elements
community groups advocated for are being stripped out of the project, and the
City of Miami government has been simply bypassing mandated review and
approval from a Citizens' Oversight Board voters approved to oversee the city
government's use of Miami Forever Bond. Nonetheless, these examples both show
how local organizations with long histories of combatting racial, economic and
environmental inequalities have strategically engaged with resilience to combat
multiple forms of insecurity and vulnerability. But importantly, in both cases, their
engagements with resilience does *not* hinge on adopting the complex ontology and
cybernetic management ethos that essentializing critiques of resilience identify as
foundational feature of resilience thinking. Instead, resilience becomes one means
amongst others to achieving particular desired ends.

To borrow the language of Giorgio Agamben (2014), resilience in these cases is
less a seamlessly operating technique of neoliberal rule than a *destituted* tool that
allows activists to seize momentary advantages that have otherwise not been available to them. For Agamben, the concept of *destitution* offers a corrective to the
dialectic between, on the one hand, an ontologically prior, affective *constitutive*
power that operates on an affective level to draw existents into world-forming
relations with one another, and on the other hand, a derivative *constituted* power
that segments and divides, vampirically leeching off the vital, world-forming force
of constitutive power (Braun and Wakefield 2018). While Agamben's (1998) earlier work on *homo sacer* essentially ontologizes the sovereign exception – that is,
offers an understanding of human life founded on the ever-present relation of the
exception (more on this below; see Coleman and Grove 2009) – his sense of destituent power offers a more contextually nuanced approach to the relation between
life (and non-life) and power. Agamben (2009) shares Baudrillard's thanatopolitical
scepticism of technology, but he locates the de-personalizing and instrumentalizing
tendency Baudrillard finds in the Bauhaus within the very structure of human
being itself. For Agamben, the use of language places the human on the slippery
slope to a de-Natured humanity. Language, in his formulation, is one apparatus
amongst others that dissociates human being from its immanent immersion in a
surrounding milieu. In his reading, the reflexive grasping of the self as the subject
of language – the 'I' – inscribes the structure of the sovereign exception – the
decision what forms of life may be valued and which may be extinguished – into
the most basic attributes of human sociality (Agamben 1978). To put this in Baudrillard's terms, the path to a de-Natured humanity begins for Agamben with the

adoption of language, and becomes intensified in modern communication technology that fulfils this de-Naturing promise by reducing the human to nothing other than one data signal amongst others, a series of digitally mediated 'likes,' 'dislikes,' purchases, searches and so forth, which de-personalize consciousness itself and open consciousness to new modes of algorithmic governance.

For Agamben, the problem here lies in the way any apparatus fabricates objects given to instrumental use (Agamben 2009, 2016). Whether this is a pencil, language or a cellular phone (Agamben 2009), the effect of an apparatus is to create a world where the human becomes an instrumental object, useful for any possible political economic project, even the catastrophe of genocidal mass killing. While Agamben may thus offer an overly confident and overly totalizing view of the exception as overcoding any and every technologically mediated interaction (Derrida 2009), his understanding of apparatus nonetheless offers a provocative vision of the politics of technologically mediated life. Here, two key concepts are *operativity* and *inoperativity*. Operativity refers to the instrumentalizing effects of any apparatus: the extent to which an apparatus renders someone or something instrumentally useful to another's goals. Operativity thus indexes an instrumental relational series, subject-apparatus-object in which the potentiality of the object (whether human or non-human, living or non-living) has been wholly actualized, and thus negated, through its becoming-instrumentalized. This effectively closes down the open-ended future of human being (Dasein), constricting this indeterminate potentiality to the determined possibilities of technologically mediated life. Thus, for Agamben, operativity signals the point where the abstract exception touches down on contextually specific, embodied practice. And liberal governance is thus sustained, on this account, through a variety of apparatuses that render all existents potentially objects of instrumental use.

This drive to operativity animates resilience ecologists' and other proponents' de-Naturing affirmations of hierarchically organized complexity, detailed in the previous section. The variety of techniques and simulation models resilience advocates use to reconceptualize existents as, inter alia, vulnerable systems exposed to long-term stresses and short-term shocks, adaptive resources, systemic threats, and so forth, render these existents amenable to cybernetic regulation and adaptive management. And yet, at this moment of what we might think, after Agamben, as the *maximum application* of a designerly, functional aesthetics to human-environment relations, the empirical examples from New York City and Miami detailed above, as well as other cases outside urban North American and European contexts (Pugh 2017), suggest the *impossibility* of a totalizing cybernetic will to design. That is, at the point where resilience thinking appears to claim a unique and singular purchase on knowing and managing change, activists and practitioners are mobilizing the concept in political projects that are *irreducible* to cybernetic governance – even if these mobilizations may fail to generate desired political outcomes (Pugh 2017; Pugh and Grove 2018).

More than clear, operationalizable policy goal that seamlessly weaves together complex social and ecological systems to protect against future shocks and stresses, resilience has also become one tool amongst others actors mobilize in a variety of situated struggles against the experiences of modernity's violence, such as the racialized violence of segregated development and exclusionary governance that has structured local economic development throughout Miami's history (Grove et al. 2018). In Agamben's terms, resilience in these examples becomes *profane*: its meaning and use are not wholly determined through its mobilization within political projects revolving around the cybernetic compulsion to reflexively manage emergent social and environmental dynamics. Profanation destitutes the determined series of cybernetic possibilities resilience offers – to be or not to be resilient, to suffer systemic collapse or facilitate systemic adaptation – and returns resilience to a more indeterminate sphere of immanent potential. Resilience becomes *inoperative,* it can be actualized in *any number* of ways, through its use within any number of political projects: anti-racist fair housing activism; shoring up fragile real estate markets; climate justice movements; maintaining urban competitiveness in a global economy; struggles to realize the promise of anti-colonial nationalism (Pugh 2017) – *all* of these projects are being advanced, with more or less success, in part through the language, imagery, tools and techniques of resilience, even as these projects are incommensurable with one another.

Importantly, the 'use' of resilience in these cases is not reducible to modernist fantasies of instrumental use that assume a pre-formed subject with clear interests. Rather, the destitution of resilience signals a kind of middle ground that mediates across subject and world – forming subjects *as* it intervenes in and attempts to shape worlds (Wakefield 2017). The inoperativity of resilience thus produces a number of paradoxical and apparently contradictory empirical effects. From community activists aligning with long-time foes in city governments to resilience initiatives that both recalibrate *and* challenge histories of racialized development at one and the same time, the newfound interest in building resilience is producing numerous inconsistencies.

Thus, the challenge for critical thought on resilience and the Anthropocene is not only to historicize celebratory affirmations of complexity, interconnection and emergence. Faced with the empirical messiness of resilience, essentializing forms of analysis attempt to show the 'truth' of resilience in ways that write out the concept's empirical inoperativity. Whether this is resilience advocates' assertions that resilience approaches offer *the* solution to knowing and managing complex challenges in today's world or critics' assertions that resilience is a fatally compromised concept that does nothing but shore up the ongoing neoliberalization of socio-ecological relations, both forms of analysis assume that resilience seamlessly transforms socio-ecological relations around the imperatives of knowing and managing complexity. In contrast, focusing on the concept's inoperativity foregrounds the indeterminate and essentially contested nature of resilience initiatives in a way that recognizes rather than disavows these inconsistencies. It helps draw attention to what Clive Barnett (2017) refers to as 'situations of practice' in which actors take up essentially contested concepts such as resilience – along with other tools and instruments – to contest specific situations of harm, insecurity or threat.

The point here is not to offer a redemptive vision of resilience that naively strives to redeem the concept for radical politics in the face of clear examples of its depoliticizing tendencies (Grove 2019). Instead, the challenge is to devise ways for critical scholars to analyse resilience that recognize rather than disavow equally clear cases of multiple (and often contradictory) political projects contesting a variety of forms of Anthropocene insecurities through, in part, the language, imagery and techniques of resilience (see also Anderson 2015). Similarly, the point is not to identify how resilience has become destituted, for this search for origins would simply repeat the essentializing manoeuvres of both critics and applied scholars that blind their analyses to multiple deployments of resilience. Such a move would implicitly position critical scholars in a privileged analytical and political position – as if the inoperativity of resilience could be effected through rarefied scholarly analysis. Instead, what is at stake here is how historicizing concepts such as resilience and the Anthropocene might destitute *forms of critique* itself (see also Collier 2017; Grove 2018). Whether these are celebratory affirmations of the Anthropocene and the vital powers of life itself or dismissive critiques of resilience initiatives, the challenge is to explore how critical thought might become inoperative and rendered useful in a variety of situated struggles against multiple experiences of insecurity, suffering and vulnerability. This holds out the possibility for critique to acknowledge rather than disavow the countless quasi-events (Povinelli 2011) of violence that sustain the Anthropocene, and the multiple ways people struggle against these micro-violences in their everyday practice (Rhiney 2019).

Conclusions

This chapter has detailed how historicizing the current scholarly interest in ontologizing 'life itself' and reifying resilience draws out a tendency within thought on both the Anthropocene and resilience to intensify rather than destabilize the modern subject's drive to eradicate alterity – either through the immunological incorporation of the other into self or through the extermination of that marked as other. While both the Anthropocene and resilience hold out the promise of a radical break with the Cartesian subject and the series of binaries that structure this subject's existence, in practice both attempt to construct a worldless humanity, a humanity whose worldly existence has been replaced by data streams and functional, systemic environments amenable to reflexive, cybernetic modulation. The affirmative politics of life and sustainability becomes an intensified politics of death that rushes to complete the modernist de-coupling of humanity and world. And yet, possibilities for contestation still emerge at the point where this politics approaches its fulfilment. As Anthropocene imaginaries de-Nature the world, the practical problem of how exactly to become resilient within particular times and spaces of the Anthropocene opens a tenuous gap that allows the violence, insecurities and injustices that created the Anthropocene to become objects of political and ethical concern. Of course, any gains social and environmental justice advocates might win through strategic deployments of resilience are at best precarious, and always on the verge of being rolled back (Pugh 2017; Rhiney 2019; Collier et al. 2017). But if resilience and the

Anthropocene are to offer any political potential, it will be because of the way they allow critical scholars and activists to draw attention to what Kevon Rhiney (2019) describes as the ongoing – and often hidden from view – struggle of marginalized folks in both the global South and North to live through and combat the everyday violence and insecurities that gave rise to the Anthropocene and the demand to become resilient in the first place.

References

Agamben, G. (1978) *Infancy and History: The Destruction of Experience*. Minneapolis: University of Minnesota Press.

Agamben, G. (1998) *Homo Sacer*. Stanford: Stanford University Press.

Agamben, G. (2009) What is an apparatus? In *What is an Apparatus? And Other Essays*. Stanford: Stanford University Press, pp. 1–25.

Agamben, G. (2014) What is destituent power? *Environment and Planning D: Society and Space* 32: 65–74.

Agamben, G. (2016) *The Use of Bodies*. Stanford: Stanford University Press.

Amin, A. and Thrift, N. (2017) *Seeing Like a City*. London: Polity.

Anderson, B. (2015) What kind of thing is resilience? *Politics* 35(1): 60–66.

Barnett, C. (2017) *The Priority of Injustice: Locating Democracy in Critical Theory*. Athens: University of Georgia Press.

Baudrillard, J. (1981) *For a Critique of the Political Economy of the Sign*. New York: Telos Press.

Bennett, J. (2010) *Vibrant Matter: A Political Ecology of Things*. Durham: Duke University Press.

Braun, B. and Wakefield, S. (2018) Destituent power and common use: reading Agamben in the Anthropocene. In M. Coleman and J. Agnew (eds.), *Handbook on the Geographies of Power*. Cheltenham: Edwin Elgar Press, pp. 259–272.

Chandler, D. (2013) Resilience and the autotelic subject: toward a critique of the societalization of security. *International Political Sociology* 7: 210–226.

Chandler, D. (2014) *Resilience: The Governance of Complexity*. Abingdon: Routledge.

Chandler, D. (2016) Securing the Anthropocene? International policy experiments in digital hacktivism: a case study of Jakarta. *Security Dialogue* 48(2): 113–130.

Chandler, D. (2018) *Ontopolitics in the Anthropocene: An Introduction to Mapping, Sensing, Hacking*. Abingdon: Routledge.

Clark, N. (2010) *Inhuman Nature: Sociable Life on a Dynamic Planet*. London: Sage.

Clark, N. and Yusoff, K. (2017) Geosocial formations and the Anthropocene. *Theory, Culture & Society* 34(2–3): 3–23.

Cohen, D. (2016) Interviews with Rebuild by Design's working group of experts. *Public Culture* 28(2): 317–350.

Colebrook, C. (2012) Not symbiosis, not now: why anthropogeneic change is not really human. *Oxford Literary Review* 34(2): 185–209.

Coleman, M. and Grove, K. (2009) Biopolitics, biopower, and the 'return' of sovereignty. *Environment and Planning D: Society and Space* 27: 489–507.

Collier, S. (2009) Topologies of power: Foucault's analysis of political government beyond 'governmentality'. *Theory, Culture & Society* 26(6): 78–108.

Collier, S. (2017) Neoliberalism and rule by experts. In V. Higgins and W. Larner (eds.), *Assembling Neoliberalism: Expertise, Practices, Subjects*. New York: Palgrave.

Collier, S., Cox, S. and Grove, K. (2017) Rebuilding by design in post-Sandy New York. *Limn* 7: 8–15.

Crowther-Heyck, H. (2005) *Herbert A. Simon: The Bounds of Reason in Modern America*. Baltimore: The Johns Hopkins University Press.

Dalby, S. (2013) Biopolitics and climate security in the Anthropocene. *Geoforum* 49: 184–192.

Derrida, J. (2009) *The Beast and the Sovereign, Volume I*. Chicago: University of Chicago Press.

Evans, B. and Reid, J. (2014) *Resilient Life: The Art of Living Dangerously*. London: Polity.

Foster, H. (2002) *Design and Crime (And Other Diatribes)*. London: Verso.

Foucault, M. (2003) *'Society Must be Defended': Lectures at the College de France, 1975–1976*. New York: Picador.

Grosz, E. (2011) *Becoming Undone: Darwinian Reflections on Life, Politics and Art*. Durham: Duke University Press.

Grove, K. (2018) *Resilience*. Abingdon: Routledge.

Grove, K. (2019) Critique, genealogy, recuperation. *Dialogues in Human Geography* 9(2): 201–204.

Grove, K. and Chandler, D. (2017) Resilience and the Anthropocene: the stakes of 'renaturalizing' politics. *Resilience: International Policies, Practices and Discourses* 5(2): 79–91.

Grove, K., Cox, S. and Barnett, A. (2020) Racializing resilience: assemblage, critique and contested futures in Greater Miami resilience planning. Accepted manuscript in press at *Annals of the American Association of Geographers*.

Gunderson, L. and Holling, C. (eds.) (2002) *Panarchy: Understanding Transformations in Human and Natural Systems*. Washington, DC: Island.

Holling, C. S. (ed.) (1978) *Adaptive Environmental Assessment and Management*. New York: Wiley.

Holling, C. S. (1995) What barriers? What bridges? In L. Gunderson, C. S. Holling and S. Light (eds.), *Barriers and Bridges to the Renewal of Ecosystems and Institutions*. New York: Columbia University Press, pp. 3–34.

Holling, C. S. (2001) Understanding the complexity of economic, ecological, and social systems. *Ecosystems* 4: 390–405.

Holling, C. S. and Sanderson, G. (1996) Dynamics of (dis)harmony in ecological and social systems. In S. Hanna, C. Folke and K-G. Maler (eds.), *Rights to Nature: Ecological, Economic, Cultural, and Political Principles of Institutions for the Environment*. Washington, D.C.: Island Press, pp. 57–86.

Huffer, L. (2015) Foucault's fossils: life itself and the return to nature in feminist philosophy. *Foucault Studies* 20: 122–141.

Jones, R., Pykett, J. and Whitehead, M. (2013) *Changing Behaviours: On the Rise of the Psychological State*. Cheltenham: Edwin Elgar Press.

Latour, B. (2004) *Politics of Nature: How to Bring the Sciences into Democracy*. Cambridge: Harvard University Press.

Latour, B. (2017) *Facing Gaia: Eight Lectures on the New Climatic Regime*. Cambridge: Polity Press.

Lee, K. (1993) *Compass and Gyroscope: Integrating Science and Politics for the Environment*. Washington, DC: Island Press.

MacKinnon, D. and Derickson, K. (2012) From resilience to resourcefulness: a critique of resilience policy and activism. *Progress in Human Geography* 37(2): 253–270.

McKittrick, K. (2013) Plantation futures. *Small Axe* 17(3(42)): 1–15.

Newell, A. and Simon, H. (1972) *Human Problem Solving*. Englewood Cliffs, NJ: Prentice-Hall.

Neyrat, F. (2019) *The Unconstructable Earth: An Ecology of Separation*. New York: Fordham University Press.

Povinelli, E. (2011) *Economies of Abandonment: Social Belonging and Endurance in Late Liberalism*. Durham: Duke University Press.

Povinelli, E. (2016) *Geontologies: A Requiem for Late Liberalism*. Durham: Duke University Press.

Pugh, J. (2017) Postcolonial development, (non)sovereignty and affect: living on in the wake of Caribbean political independence. *Antipode* 49(4): 867–882.

Pugh, J. and Grove, K. (2018) Assemblage, transversality and participation in the neoliberal university. *Environment and Planning D: Society and Space* 35(6): 1134–1152.

Rhiney, K. (2019) Rethinking resilience and its ethico-political possibilities. *Dialogues in Human Geography* 9(2): 197–200.

Rose, C. (2015) *The Struggle for Black Freedom in Miami: Civil Rights and America's Tourist Paradise, 1896–1968.* Baton Rouge: LSU Press.

Sexton, J. (2011) The social life of social death: on afro-pessimism and black optimism. *InTensions Journal* 5: 1–47.

Simon, H. (1955) A behavioral model of rational choice. *Quarterly Journal of Economics* 69(1): 99–118.

Simon, H. (1956) Rational choice and the structure of the environment. *Psychological Review* 63(2): 129–138.

Simon, H. (1962) The architecture of complexity. *Proceedings of the American Philosophical Society* 106(6): 467–482.

Steinberg, P. and Peters, K. (2015) Wet ontologies, fluid spaces: giving depth to volume through oceanic thinking. *Environment and Planning D: Society and Space* 33(2): 247–264.

Stiegler, B. (2018) *The Neganthropocene.* London: Open Humanities Press.

Thomas, D. (2016) Time and the otherwise: plantations, garrisons and being human in the Caribbean. *Anthropological Theory* 16(2–3): 177–200.

Tsing, A. (2015) *The Mushroom at the End of the World.* Princeton: Princeton University Press.

Wakefield, S. (2017) Inhabiting the Anthropocene back loop. *Resilience: International Policies, Practices, Discourses* 6(2): 77–94.

Walker, J. and Cooper, M. (2011) Genealogies of resilience: from systems ecology to the political economy of crisis adaptation. *Security Dialogue,* 42(2): 143–160.

Watts, M. (2015) Now and then: the origins of political ecology and the rebirth of adaptation as a form of thought. In G. Bridge, J. McCarthy and T. Perreault (eds.), *The Routledge Handbook of Political Ecology.* London: Routledge, pp. 19–50.

Yusoff, K. (2013) Geologic life: prehistory, climate, futures in the Anthropocene. *Environment and Planning D: Society and Space* 31(5): 779–795.

Zebrowski, C. (2016) *The Value of Resilience: Securing Life in the Twenty-First Century.* Abingdon: Routledge.

8

IRONIES OF THE ANTHROPOCENE

Lauren Rickards

Introduction

Among the many practical problems the Anthropocene presents is the question of how to orient ourselves to it, intellectually and institutionally. For academics, especially, working out how dark the Anthropocene story is, how deeply to absorb its unsettling pronouncements, and how to respond are acute professional as well as personal challenges. Is the Anthropocene, for instance, a ridiculous melodramatic flourish, a mere Hollywood-esque hunger for drama? Is it even a dangerous idea, a new seductive armoury for misanthropic or hyper-anthropic antics, that we would be doing a mis-service to our profession and species if we were to allow it into the academy? Some stratigraphers are highly suspicious of the concept (e.g. Rull 2013; Autin and Holbrook 2012), while anyone giving a moment's thought to the name of the new self-declared epoch is likely to raise an eyebrow if told its job is to instil a new environmental ethic (e.g. Crist 2013). But is the risk, instead, that we dismiss the Anthropocene's vital message out of dislike of its clumsy name and apparent pretensions? Is the real danger, as Ian Angus (2016) asserts, that we ignore the clarion call of the Anthropocene or pause at the very moment we need to jump in and tackle our civilisational catastrophe head on? Certainly there is a growing litany of reasons for sitting up and paying closer attention to the destruction unfolding around us.

As academics, these perplexities are far from incidental. The Anthropocene seems to give new meaning to the term 'flexible knowledge workers' (Swan and Fox 2009), calling us to push our hands into the soil of real world practice at the same time as we lift our eyes to the starry skies of cosmological and existential questions. It also means being a knowledge worker at a time when to know is to feel increasingly unhinged from any certain knowledge. Any strategy that promises us to help manage the resultant strains warrants close consideration.

In this chapter I explore one such option – an 'ironic world relation' (Szerszynski 2007) – based on dwelling consciously within rather than clamouring over the perplexities outlined above. In general, irony refers to 'a condition of affairs or events of a character opposite of what was or might naturally be expected; a contradictory outcome of events as if in mockery of the promise or fitness of things' (Oxford English Dictionary 2015). As discussed below, the emergence of the Anthropocene can be read as itself an irony or unintended consequence. As Dana Phillips puts it, the Anthropocene is:

> An ironic term: it sounds triumphant while being anything but; it signals that we have reshaped the planet to suit ourselves, and acknowledges that we have done a great deal—perhaps even a fatal amount—of harm along the way.
>
> *(Phillips 2015 : 6)*

Such a realisation however does not explain the further irony of the ongoing perpetuation of the Anthropocene. The very fact that fossil fuel extraction, for instance, is still being pursued by some actors in the age of the Anthropocene when its illogicality is blindingly apparent illustrates the sort of recursive, multi-layered knowledge politics the Anthropocene involves. To a significant degree, these messy politics stem from the inadequacies of the easy, early intellectual responses to the Anthropocene that critical social scientists have been among the advocates of. They underline the perplexing ironies of the new epoch and the need to move beyond familiar intellectual maps and moves.

In the sections below I focus in turn on some of the communicative and situational ironies posed by the social phenomenon of the Anthropocene. In doing so, I engage with a broader, increasingly popular, strategy for dealing with climate change and the multiple catastrophes represented by the Anthropocene: humour. As a growing number of scholars argue (e.g. Fluri and Clark 2019; Ridanpää 2014), the darker the present and future seems to become, the more seriously we need to take humour as an intellectual and affective response. I then move beyond communicative and situational ironies to discuss the potential of Szerszynski's (2007) idea of an ironic 'world relation' as an encapsulation of the sort of orientation the Anthropocene requires. First, though, I begin by outlining the multifaceted relationship of irony to environmental crises.

Irony as trouble or tool?

Writing in 2007, German political theorist Ingolfur Blühodrn noted that: 'There is a striking consensus between political elites and general electorates that it is time to stop *talking* about things and take *decisive action*: Cut though the rhetoric! Get down to the issues!' (Blühdorn 2007: 252). Blühodrn is cynical about these intense exhortations to 'do something'. Following Nullmeier (2005), he asserts that such cries are empty rhetoric, a mere 'performance of seriousness' serving to spin the wheel of 'simulation politics'. In the latter, 'contemporary culture and the meanings through which it is reproduced have become self-referential, detached from any obligation to material referents or effects' (Szerszynski 2007: 338). The twin outcomes of such illusory efforts are deleterious real world impacts and a 'post-ecological' condition. In the latter, worsening environmental troubles only fuel circular discussions, technocratic responses and justification of their ongoing amplification in states that resemble the 'post-political' stupor that Eric Swyngedouw warns of (Swyngedouw 2010; 2013; 2007) or the 'crisis in political meaning' that Bronislaw Szerszynski (2007: 338) refers to. The result is what Blühdorn (2011) calls the 'ecological paradox': knowing what is going on but doing nothing anyway.

In such light, the Anthropocene can be read as cause and/or effect: a long period of creating, justifying and ineffectively treating environmental degradation; and the name and symbol of the current, consequent planetary crisis, that a substantial proportion of those who know about it still ignore. As the latter, the question is whether its recent naming simply folds into the rhetorical haze of the former – adding to an ongoing 'performance of seriousnesss' as it ratchets up and freeze-frames environmentalist distress – or whether it is in fact a long-awaited opportunity to break out of the attendant crisis in political meaning and commence a new period of effective collective action.

Like Blühdorn, sociologist Bronislaw Szerszynski believes in the reality and urgency of the ecological problems encompassed by the Anthropocene, but he is not as pessimistic about the potential for change. Rather, he holds out hope that we may be able to escape from our current political paralysis by adopting a high-level stance of irony (discussed below). He is not alone in turning to irony. Ridanpää (2009) suggests that irony is crucial as 'a tool through which the geopolitical order and geopolitical meanings become negotiated and social self-identities established' (2009: 730).

That scholars are suggesting irony as a useful and action-enabling attitude may seem ironic in itself. For, although irony is an ancient form of rhetoric and 'has a long tradition of being a rather potent form of social commentary' (DiCaglio 2014: 2), it is also a 'risky business', functioning 'tactically in the service of a wide range of political positions, legitimizing or undercutting a wide variety of interests' (Hutcheon 1994: 10). Furthermore, it is a product of (cultural) modernism of the sort blamed for the Anthropocene, and is widely criticised as a postmodern indulgence, dismissed as an unhelpful source of apathy, disengagement and scorn. Karlsson (2013), for instance, presents an ironic stance to the Anthropocene as an easy, self-serving cop-out:

> When confronted by, on the one hand, the unsustainable nature of existing socio-economic arrangements and, on the other, the radicalism of any meaningful alternative, one possible response is to retreat into post-modern irony. As all irony, it is a stance which requires minimal personal engagement. Instead of taking active responsibility for the future and trying to articulate intelligent ways of moving society forward, such an ironic stance is

often characterized by apathy and resignation … Instead of idealism we find
a growing cynicism, a cynicism which in itself is then often used to prove the
impossibility of idealism.

(Karlsson 2013: 4)

What Karlsson is referring to is 'dispositional irony': a strategic self-conscious mode
of comportment characterised by cool observation of contradictions between
others' beliefs and circumstances. Such a stance can be highly negative. It is what
Timothy Morton gestures to when he says postmodernism 'freezes irony into an
aesthetic pose' (Morton 2007: 21). But as discussed below, Morton, like Szerszynski, also points to the value of irony in addressing the ecological crisis we face
(e.g. Morton 2010). For, in more general terms, an ironist is an 'explorer of
incongruities' (Donoghue 2014: 138), and it is incongruities that the Anthropocene
throws up. As Gibbs argues:

Irony is among our best methods for immediately and unconsciously adjusting
to complex circumstances. Embracing irony allows us to cope with the dis-
parities we experience and express something about the inchoateness of the
human condition. Perhaps, ironically, irony may be the best verbal form for
expressing what we most earnestly believe.

(Gibbs 2002: 152)

Others make similar suggestions about dark humour more generally. On the one
hand, dark humour can perpetuate and mask a sense of helplessness and bleakness,
and be wielded in an exclusionary and offensive way. On the other hand, it can
serve as a 'survival mechanism', 'tool of resistance' and source of comprehension
(Eriksen 2019; Fluri and Clark 2019: 123). The question then is not whether we
should engage with humour and irony specifically, but how.

Ironies of the Anthropocene

Communicative ironies

There are numerous ironies embedded in the Anthropocene, including what
Szerszynski calls 'communicative ironies' where 'the overt, surface meaning of the
communication is in tension with the actual meaning intended to be commu-
nicated' (Szerszynski 2007: 341). The primary example of this is the juxtaposition
between the Anthropocene as an intended warning and its reception in various
quarters. Contradicting the assumed trajectory of human society, the mainstream
scientific Anthropocene narrative turns the dominant Progress Narrative of Western civilisation on its head and exposes presumed improvement to be actually a
Declentionist Narrative of decline (Whitehead 2014).[1] As Castree notes, numerous
scientists associated with the Anthropocene thesis 'have used their institutional
authority to sound the environmental alarm louder than at any time since the early

1970s', speaking out 'from respected universities within the heartlands of political economic power' (Castree 2015: 55). Although it is highly debatable how unexpected or even unintended the attendant environmental degradation is, or how helpful a Declentionist reading of the situation is (Merchant 2003), the implicit cultural expectation that human society has been improving upon and insulating itself from nature means that the recent Anthropocene report on our *actual* negative, life-endangering effects has a decidedly ironic tone, making it a case of the 'situational ironies' (gap between expected and actual reality) described below. A humorous representation of this general situation is provided in an episode in Season 3 of the US NBC comedy series *The Good Place*, where it is discovered that, despite lifetimes of do-gooding, no human has entered Heaven for 521 years, the reason being that the indirect consequences of these actions were negative, doing greater total harm to the environment.

The focus here is how, despite or perhaps because of the seriousness of its tone and message, the Anthropocene thesis has been met with splutters and sneers among some audiences, including some critical social scientists (Castree 2014; Castree 2015). That anyone could purport to know the whole Earth, diagnose a new condition, pinpoint causes and prescribe solutions seems laughably hubristic. Innumerable social science critiques have pointed out the 'god trick' (*cf* Haraway 1988: 582) that such thinking and its presumption of infinite vision relies on, and the highly partial positionality (elite, white, male scientist) that the trick tries to disguise. Writing about the use of images of the Earth-from-space in public Anthropocene discourse (e.g. the Anthropocene exhibit at the Haus der Kulturen der Welt (HKW) in Berlin), Lekan, for example, argues that:

> the attempt to depict *Anthropos* as a unitary geophysical agent resurrects the appeal to the Whole Earth environmentalism of the 1970s without attending to the U.S. imperialist and racist connotations of the disembodied 'god trick' found in these extraterrestrial photographs.
>
> *(Lekan 2014: 171)*

This and similar critiques of the Anthropocene thesis (e.g. Crist 2013; Lövbrand et al. 2015; Swyngedouw and Ernstson 2018) are enabled not just by the global scale and extreme ambition of Anthropocene science, but by the honing of the underlying basic critique of such science and environmental messages over past decades thanks to the hard work of feminist, postcolonial, political ecology scholars and systems thinkers, including critics of conventional Natural Resource Management (discussed below). Another irony, then, of the Anthropocene is the juxtaposition between, on the one hand, the ambition and solemnity of its scientific claims about the endangered Earth, and, on the other hand, the scoffing such claims have triggered among the many members of the audience arguably concerned about environmental degradation but highly sensitised to the perversities of top-down declarations.

Moreover, confident scientific assertions about understanding the world and how to (now, better) 'Command and Control' it are themselves frequently blamed for not just diagnosing but contributing to the environmental disruption of the Anthropocene. In other words, not just the Anthropocene condition but the authority of Anthropocene science is, ironically, enrolled as evidence of the need for further reflexive modernisation of the sort diagnosed by earlier theorists. For example, in his work on the governance challenges posed by the Anthropocene, Jonathan Pickering highlights the relevance of Anthony Gidden's work on 'the paradox that modern approaches to governance have aspired to technocratic control, yet in doing so have crafted institutions and processes that yield further instability and uncertainty' (Pickering 2018: 4), while Anna Volkmar (2017) notes that the Anthropocene reflects Ulrich Beck's (1992) idea of modernity as a newly uncertain, reflexive stage, the age of 'unintended consequences'. Others argue that escaping the domination by science and technology is not just a matter of avoiding poor decision making, it is a matter of empowering other ways of thinking and acting, including those excluded from the universal figure of the Human that science relies upon. As Luisi Pellizzoni (2014) writes about Max Horkheimer and Theodor Adorno's influential book *Dialectic of Enlightenment*,

> it detects in modern reason a tragic contradiction between human attempts to find protection from nature and the unknown through knowledge and technical control, and subjection to these very instruments of emancipation. Getting free from the hold of nature means falling prey to the same natural forces of domination.
>
> *(Pellizzoni 2014: 198–9)*

Given all of this, the irony of the top-down scientific nature of the Anthropocene declaration by a small group of elites is greater than the fact that it has failed to resonate with some critics; it is that its form and source seems to deny what many argue has contributed significantly to the Anthropocene problem in the first place.

A particularly 'ferocious' form of irony is satire (Donoghue 2014: 138) and a host of it has arisen in response to the Anthropocene and its component issues such as climate change. Boykoff and Osnes (2019) note that satire and other forms of comedy offer a valuable, if unpredictable, means of positioning climate change within the politics of everyday life, helping people engage with the issue in a way that scientific information and formal pronouncements often fail to do. Political satire can especially expose contours of self-interest, contradiction and irony, undermining the power relations involved (Kuus 2008). While video material is increasingly dominant, political cartoons and comic strips remain an especially punchy, powerful medium, serving not simply to 'jazz up' information about environmental issues but to stimulate thinking about the politics and acts of representation involved (e.g. Manzo 2012). As Robson (2019: 115) notes, because such images and narratives 'lean so heavily on the figurative' or 'appeal to humour', 'their political verve can sometimes be underestimated or dismissed', giving them a subversive potential. This is especially the case with comic strips which have been conventionally dismissed as 'low brow' culture and pure entertainment (Dittmer 2005).

A classic example is the work of popular UK artist Marc Roberts (cartoonist with the *New Internationalist*), who frames his potent critiques of the status quo as 'dark green humour to brighten your day'.[2] Aimed at far more than eliciting a laugh, Roberts' comic strips utilise communicative irony purposively and productively. In particular, Roberts' 'Anthropocene' comics play on not only the irony of the Anthropocene as an unintended side-effect of human progress, but the irony that some people (especially those most committed to the Progress Narrative and techno-optimists) do not hear or accept the profound critique of industrialisation encompassed within scientists' Anthropocene message and instead perversely misinterpret the Anthropocene announcement as an *endorsement* of current trajectories. One comic strip, for instance, depicts a suited white male punching the air with satisfaction and shouting 'Go humans!' at the naming of the new epoch (Figure 8.1). It presents as comic society's conventionally self-reassuring and technocratic responses to serious signs of environmental degradation. Techno-ecological modernist confidence is the target of Roberts' entire 'Frank' series that the strip is part of. Its older character – 'Cantankerous Frank' – rails against the contradictions, frustrations and self-centredness of the smooth-talking 'Government Environment spokesman' Ernest Readham – the man in the suit – who in the first cartoon of the series Frank refers to as a 'mealy-mouthed, ladder-scrambling, piss-puddle of complacency'.[3]

For most of us interested in the Anthropocene, this comic strip will probably induce smiles, snorts and grimaces, illustrating the way cartoons can cleverly capture the comic contradictions within issues. Besides providing ever-more-important light relief and facilitating our mental resilience in the face of the Anthropocene and related crises (Ridanpää 2019), such humour can be a valuable tool in identifying issues of analytical importance and helping us confront what is otherwise hard to comprehend, represent or accept (Carty and Musharbash 2008; Ridanpää 2009), as literary scholarship on climate change and the Anthropocene indicates (e.g. Shewry 2019). Humour also often functions to create rapport with like-minded others and thus to shore up our identity. As Colebrook (2000) notes, the success of ironic speech acts relies on the identity of the speaker being recognised and stable.

FIGURE 8.1 'Anthropocene' comic strip by Marc Roberts as part of the 'Frank' series[4]

Resultant social connections can help fuel vital resistance against hegemonic parties (Ericksen 2019; Ridanpää 2010).

However, the use of irony and satire can also foster exclusion and aggression (McCullough 2008, Ridanpää 2019). Karlsson (as noted above) points out that irony can quickly degenerate from critical observation into smirking, self-serving mockery of the deluded. Even if such excesses are avoided, communicative irony is often based on a sense of 'insiders' who get the joke and 'outsiders' who do not:

> Irony is a form of utterance that postulates a double audience, consisting of one party that hearing shall hear and shall not understand, and another party that, when more is meant than meets the ear, is aware both of that more and of the outsiders' incomprehension.
>
> *(Fowler 1923 in Donoghue 2014: 138)*

Thus, while speaking at multiple levels through irony can be a valuable rhetorical strategy, it invokes multiple risks. As Fluri and Clark (2019: 124) note, 'no dominating power is immune from laughter' and this includes those of us who use communicative irony.

The first risk is that, as Fowler indicates, irony can create a sense of (smart) Us and (dumb) Them. This, in turn, can contribute to a polarisation of politics and its depiction of 'a struggle between "the people" and some combination of malevolent, racialized and/or unfairly disadvantaged "Others"' (Scoones et al. 2018: 2). The associated rise of authoritarian populism fuels a backlash against smirkers and knockers, often perceived as an educated, urban, greenie elite. Recent election results in Australia and United States – while the outcome of multiple factors (Monnat and Brown 2017) – demonstrate that this is not an incidental concern, contributing among other things to the dismantling of environmental regulations and generation of more Anthropocene-inducing environmental disruption. Environmental approvals for the massive, controversial Adani coal mine in Australia, for instance, were accelerated following the 2019 re-election of a far-right conservative national government, headed by an ex-marketer and largely ushered into power by disenfranchised rural and regional voters within the mine's state of Queensland.[5]

Second, communicative irony risks amplifying not just disenfranchisement and anger, but misunderstanding. Ridanpää (2009: 322) suggests that 'Irony means simply a misunderstanding of the words presented, a comprehension of them in the opposite way from that in which they were meant'. But misunderstanding may be accidental as well as intentional, triggered by an ironic use of a term as much as a conventional one. In the comic strip described above and others like it, the implied misinterpretation of the Anthropocene is not corrected, meaning that 'outsiders' may simply accept the ironic interpretation presented and, in a further ironic twist, proceed to act as if the 'Age of Humans' does indeed mean the 'Age of Human Achievement'. This is especially the case if non-ironic uses of key terms are circulating in popular discourse, as they are in the case of the 'Age of Humans' (e.g. Lynas 2011). The phrase 'Welcome to the Anthropocene' is another such example. It is now famously satirised in art by Robyn Woolston in her 2013 Habitus installation[6] (Figure 8.2), which, with its tongue-in-cheek insertion of the word 'fabulous' and 1950s aesthetic, elegantly riffs off the favoured post-World War II start date of the Anthropocene (see Steffen et al. 2016). Despite the irony of the word 'Welcome', the phrase is used in a *non-ironic* fashion in some academic and policymaker discourse, including *The Economist* and an influential video used to open the United Nation's 2012 Rio+20 summit.[7] Clive Hamilton (2016) laments:

> Some scientists even write: 'Welcome to the Anthropocene.' At first I thought they were being ironic, but now I see they are not. And that's scary. The idea of the Anthropocene is not welcoming. It should frighten us. And scientists should present it as such.
>
> *(Hamilton 2016: 251)*

The combination of these two risks leads to a third, which is that even the most carefully crafted communicative irony is simply swept up into the mass of swirling assertions, facts, Trumpian 'alternative facts' and texts which characterise the era of social media, 'post-truth' and authoritarian populism. The resultant mess of meanings makes discerning the importance and source of messages extremely difficult. More than the risk that communicative irony is misunderstood, at work here is the

FIGURE 8.2 Welcome to the Fabulous Anthropocene Era. One of 12 pieces in the 2013 Habitus installation by Robyn Woolston, at Edge Hill University, Lancashire, UK

risk that deconstructive modes of communication, like irony, fuel far-right agendas by providing readily co-opted messages. As Neimark et al. (2019: 4) note:

> the latter have knowingly borrowed tactics and strategies used by left-leaning activists and scholars to highlight the politics of knowledge production, to push for the acceptance of alternative facts and to relativize the views of scientists and right-wing ideologues.

For critical scholars this requires confronting the 'potent conundrum' and paradox of the current political moment by 'addressing those seeking to obfuscate or deny environmental degradation and social injustice' while still critiquing 'the privileged role of Western science and expert knowledge in determining dominant forms of environmental governance' (Neimark et al. 2019: 1).

It is significant to note here the relationship to the efforts mentioned above to deconstruct Anthropocene science. The common misuse of consequent critiques layers atop scientists' ironic lack of reflexivity the further irony of the feeding of far-right anti-environment interests. For, as Haraway (1988) notes, the strong constructivist programme of postmodernism went beyond 'separating the good scientific sheep from the bad goats of bias and misuse' to unmasking 'the doctrines of objectivity'. But the subsequent abandonment of objectivity has created difficulties for the former job of sorting through information, creating problems: 'for those of us who would still like to talk about reality with more confidence than we allow to the Christian Right when they discuss the Second Coming and their being raptured out of the final destruction of the world' (Haraway 1988: 577).

In the two decades since Haraway wrote this, these difficulties have grown thanks to the need to now ensure that such interests do not just jam the airways with trash talk, but actively seize upon scientific and ironic messages (such as those now abounding about the Anthropocene) to support their own ends, including techno-optimist narratives about the 'end of nature' (Shellenberger and Nordhaus 2004), fatalist narratives about the end of the world (Ronan 2017; Lilley et al. 2012), and neoliberal efforts to wind back scientific, legal and academic institutions on the purported basis that scientific expertise and elite knowledge are now obsolete (Mirowski 2019).

Overall, if our use of eco-satire inadvertently leads to perverse anti-environment outcomes, perhaps the joke is on us for believing in the power of communicative irony to identify needs and stimulate positive social change. It may be for this reason that in 2011 Marc Roberts suspended the Frank cartoon series, noting with some bitterness that Frank's 'nay-saying' and 'dissing' of 'the "opposition"' has created a counter-productive 'cul de sac' by not saying 'anything positive, which is what is so desperately needed in a culture obsessed with happy-ever-afters and consumable frivolity'.[8]

Situational ironies

Some scholars and environmental activists are increasingly tackling environmental inaction with a more forthright, earnest and public form of irony of the sort that

Szerszynski (2007) refers to as 'situational irony', the gap between expected or intended and *actual* situations that are 'inherent within unsustainable and unjust cultural practices' (2007: 347). Use of situational irony reflects a commitment to 'speak truth to post-truth' (Neimark et al. 2019) and to resist 'attempts by powerful elites to turn language into a mere tool to be deployed to further their own projects' (Szerszynski 2007: 347). Situational irony exists 'when the understanding of a situation possessed by one or more actors acting within that situation is in dramatic tension with the reality of it as perceived by an outside observer' (Szerszynski 2007: 341). A common aim is to publicly expose the *cynicism of others* (namely policy makers and corporates, but also peers) by pointing to the contradictions between their empty rhetoric and promises on the one hand, and their actual (in)actions and negative effects on the other. Rather than pointing out just ignorance, it calls out simulation politics and greenwash of the sort that abounds in the post-truth era, exposing it as a calculated attempt to appear to 'do something', while actually doing nothing, or worse.

Situational irony is implicit in the Frank series in that the series is premised on the disjuncture between Earnest's title as the 'Government Environment Spokesperson' and his actual anti-environment attitude. Such ironies are also common targets in much critical social science, notably political ecology, where discrepancies between the (stated) green intent and actual actions or effects of various public and private initiatives or principles are frequently highlighted. The inadequacies of ecomodernist techno-optimist policies and practices provide an endless list of examples to investigate. In an article entitled *Welcome to the Anthropocene*, Clémençon parodies the UN's Rio+20 'grand vision' for sustainable development. He argues that behind the fanfare 'Rio+20 continues to subscribe to a business-as-usual economic growth model that—despite concerns expressed to the contrary—assumes an unlimited capacity of the planet to supply nonrenewable and renewable resources and absorb global pollution' (Clémençon 2012: 311). Bailey et al. similarly examine the 'tensions, contradictions and limits' involved in the application of ecological modernist principles to climate change, which has allowed 'the seemingly intractable problem of climate change' to be rapidly 'reframed as an opportunity to construct a new carbon economy' (Bailey et al. 2011: 683). They point to the irony that, despite much talk about how profound and intractable an issue climate change is, the ecomodernist carbon market approach being taken actually bolsters the economic growth imperative that is fuelling climate change in the first place (Bailey et al. 2011: 683).

A similar issue is evident in the use of adaptive management and its underlying critique of Command and Control institutions and approaches (Holling 1978). Adaptive management – that is, managing in a way that acknowledges complexity and uncertainty and the need to learn and adapt over time – is part of what Sayre (2012) calls the 'conceptual scaffolding' of the Anthropocene. As Simpson notes, the Anthropocene is 'a bundle, knit together from already established languages and frameworks of understanding' (2018: 2). This includes an understanding of the world in systemic terms of the sort that also underpins the parallel idea of adaptive management (see Grove 2017 on the rise of the latter). Unsurprisingly, adaptive management is touted as a new improved route to environmental sustainability in the Anthropocene, including by some Anthropocene scientists. Steffen et al. (2011) suggest, for example, that: 'Active adaptive management has proven effective in dealing with complexity and uncertainty at smaller levels and might also be effective at the global level' (2011: 857). Others agree, with the Anthropocene fostering the expansion of adaptive management applications (e.g. Kelly et al. 2018; McGinley 2017). Yet, ironically, adaptive management presents the risk of deepening capitalist exploitation of the planet in three ways, one ontological and two practical.

First, underpinning adaptive management is a vision of nature as profoundly complex, non-linear and unstable. Associated with the non-equilibrium paradigm of ecology that naturalises disturbance and challenges the idea of stable climax communities (e.g. Gaud et al. 1984), as well as with earlier and parallel developments in complexity science, systems thinking and quantum physics, this conception of nature vastly multiplies its relations, extends its scales and blurs its apparent boundaries. This includes the boundary between humans and the rest of nature, challenging the dominant Cartesian conception of Human and Nature along with Cartesian notions of space and Newtonian notions of physics. The resultant image of the Earth as an interconnected complex adaptive system is the ontological basis of Earth System Science, which along with stratigraphy has authored the Anthropocene thesis. At one level, this shift in perception is strongly associated with more sustainable, resilient and adaptable approaches of the sort numerous Anthropocene authors call for. For example Steffen et al. (2018) call for us humans to think of our actions as feedbacks on the Earth System, arguing that to manage future planetary trajectories responsibly and steer the Earth away from a dangerous functional threshold, beyond which lies cascading life-threatening changes, 'human societies and our activities need to be recast as an integral, interacting component of a complex, adaptive Earth System' (Steffen et al. 2018: 8256). At another level, however, the relational systemic ontology associated with the Anthropocene converges, ironically, with a neoliberal conception of the world and risks facilitating the further marketisation and exploitation of the planet. It does this in two ways:

1. via the common intellectual roots of the science with neoliberalism, its operationalisation into a new 'resilient' model of natural resource management and the emergent applications of neoliberalism in all elements of the state (see Grove 2017; Walker and Cooper 2011); and
2. via the relational material ontologies that various humanities and social science scholars have sketched out in explicit or implicit response to the Anthropocene condition, which Swyngedouw and Ernstson (2018), Braun (2015a) and Pellizzoni (2016) assert are inadvertently helping to adapt the new scientific view for capitalist interests.

There is thus a least a partial ironic discrepancy between the hopes invested in the new scientific understanding of the world and the outcomes that new understanding is leading to.

Second, a further discrepancy emerges out of the gap between the high hopes invested in adaptive management as a route to more progressive sustainable outcomes, and how it is being applied in practice thanks to its cynical cooption by capitalist interests. Adaptive management per se arose out of C. S. Holling's intellectual critiques of what he labelled the Command and Control 'pathology' of natural resource management (Holling and Meffe 1996). Fuelled by its convergence with similar assessments of top-down approaches from critical social science and neoliberal quarters, as well as the take-off of the associated field of resilience science, this critique of Command and Control natural resource management expanded to encompass the wider world of resource governance, where an ideal of flexibility, adaptiveness and a 'learning orientation' was similarly introduced (see Folke et al. 2005; Grove 2017). One result has been the progressive incorporation of adaptive management within environmental and planning law, where it is celebrated as a more flexible, responsive, effective and often inclusive way of governing potential environmental impacts of resource-based activities (e.g. Benson and Garmestani 2011; Craig and Ruhl 2014). Rather than using rigid licences and regulatory standards that may misunderstand whether and how negative impacts emerge, legal institutions are now building in a monitoring and response process that requires developers to assess impacts over time and to adjust their activities accordingly if negative impacts become evident. In this way, a development is given only contingent permission to proceed and the potential for more restrictive 'laws of limitation' (cf Fisher 2014) is held open. The irony, however, is that the flexibility inherent to adaptive management is being used in some cases to fast-track risky developments and to legitimate environmental degradation as the indeterminacy of natural system responses is used in pre-development Environmental Impact Assessments to disguise three things: strategic ignorance; uncertainties stemming from lack of analysis of likely impacts; and the relevance of the precautionary principle (see Lee 2014). In this way, early fears that the new disturbance ecology paradigm would facilitate environmentally damaging activities are being borne out, though less as a result of anything inherent to the idea of adaptive management than as a result of its cynical application by vested interests, as Zimmerer (1994) predicted.

That brings us to the third situational irony associated with the adaptive management paradigm. One of the ways in which adaptive management is being misused/abused is that its application in environmental law is frequently not accompanied by legally binding objectives and standards or stated obligations around monitoring or adaptive responses, remaining instead a vague gesture about dealing with things 'down the track' (Lee 2014). Often contributing to and disguised by these ineffectual legal references to adaptive management is a lack of underpinning scientific information about the existing condition and character of a given natural system, or rigorous analysis of how it may be affected by the proposed intervention. In the case of the proposed Adani mine (mentioned above), for example, the hydrological models used by the mine proponents to estimate the effects of the mine on the groundwater reserves, and subsequently justify the granting to the mine of unlimited access to groundwater, have been shown to be patently inadequate. Building on Currell et al. (2017), a recent Position Paper by Concerned Scientists[9] – Werner et al. (2019) – outlines the 'deficiencies in the scientific assessment of the Carmichael Mine impacts to the Doongmabulla Springs', a significant ecological system fed by groundwater in the vicinity of the mine. It argues that:

- Adani appears likely to have significantly under-estimated future impacts to the Doongmabulla Springs Complex (DSC) arising from the Carmichael Mine.
- Should the Carmichael Mine cause springs within the DSC to cease flowing, this impact may be irreversible.
- The safeguard against DSC impacts proposed by Adani, namely Adaptive Management, is unsuitable and unlikely to protect the DSC from severe degradation or cessation of flow.
- Possible cumulative impacts to the DSC from other mining activities in the Galilee Basin have not been adequately considered.

(Werner et al. 2019: 1)

They conclude that 'the DSC face a legitimate threat of extinction due to the Carmichael Mine project' (2019: 1). There are numerous things to note here. One is the way that the scientists are not pointing out how obviously reckless granting unlimited groundwater access to a mine is in the context of other users and threats to the groundwater system in question (the Great Artesian Basin). Instead they are limited to critiquing the impacts of groundwater extraction upon a particular eco-logical system (the Doongmabulla Springs Complex) because only a limited array of environmental issues are legally valid under the national Environmental Protec-tion and Biodiversity Act. Also notable is the bracketing out of climate change as a relevant issue from the legal proceedings, despite the extreme impact the produc-tion, transportation and combustion of the coal will have on the climate. It is also important to recognise the separate elements of law in play. The state of Queens-land's 'laws of competence' are what allow it to grant access to resources to those it deems suitable. This wielding of law precedes and thus trumps the 'laws of limita-tion' that may *subsequently* mediate such access (Fisher 2014). The upshot is that the use of adaptive management in the latter presents a classic case of how bounded resilience thinking can conceal the highly political generation of disturbances in the first place. As many scholars have recognised, it is vital to ask 'resilience to what?' (e.g. Carpenter et al. 2001). Or as Braun puts it:

the implicit demand to 'adapt' or 'experiment' ignores crucial geopolitical and political economic questions of who is compelled to experiment with new ways of knowing and producing novel socio-ecological forms, and who is in the position to make the demand.

(Braun 2015b: 112)

The question of 'who is in the position to make the demand' brings us to the work of the Concerned Scientists noted above. A group of hydrologists and other geoscientists, these scientists are far from typical activists, but have been pushed to 'weaponise' (politically activate) their scientific privilege by the egregious misuse of their profession by the state to legally justify the highly destructive mine. More precisely, they have been pushed to 'counter-weaponise' their science to oppose how the corporate-state regime has already weaponised the tools and techniques of their field to support desired political projects. This move reflects a wider trend by scientists to issue dissenting scientific views in the face of environmental mismanagement based on bad (deliberately inaccurate) science. In Australia this includes recent efforts by the Academy of Science and Wentworth Group of Concerned Scientists to call out the mismanagement of the enormous Murray Darling Basin river system, where over-allocation of water and the exclusion of climate change projections from models and planning has contributed to extreme environmental degradation.[10]

Combined with the parallel instrumental use of science to underpin corporate bids for extractive uses (certain geological surveys in the case of coal mining, certain agronomic studies in the case of irrigated mega-farms in the Murray Darling Basin), and efforts to try to discredit dissenting views, this points to the variation and contestation that characterises science. Contestation is arguably especially apparent in the Anthropocene when vastly different groups are competing to present visions of the future and the transformations needed. One of the features of the so-called Great Acceleration (the radical jump in production and consumption following the Second World War) is the partial emergence of the 'environmental-social impact sciences' to contest (not very successfully) the dominant 'technological-production sciences' (Schnaiberg 1980), as illustrated by the associated rise of adaptive management. One of the physical and intellectual products of this time – anthropogenic climate change – has especially helped some scientists appreciate the stakes of their work (e.g. Tadaki et al. 2014). Climate denial attacks on climate science have demonstrated the politics such science is inescapably part of, triggering a wide range of (idealised) responses among the scientists involved (see Lewandowsky et al. 2015; Lewandowsky et al. 2013; Maibach et al. 2014; Hoggett and Randall 2018). But, ironically, such heterogeneity and conflict is papered over in most social science representations of science, which are rarely more precise than specifying a scientific discipline, if even that. As a result, important debates in, for instance, Anthropocene science – between Earth System Scientists and stratigraphers – are ignored (Rickards 2015).

Homogenising and discrediting all science in the name of progressive politics (focused on single issues, such as resilience or the Anthropocene) presents two component ironies. The first is that it neglects subjugated knowledges and voluntary activities within science to present a black-boxed view of science. Such a representation can act as a microcosm of how many (social science) narratives of the Anthropocene ironically repeat the modernist mistake of simplifying history into a linear series of homogenised eras, representing 'modernity as monolithic and total, dividing its history into an un-reflexive (pre-)modernity and a post-evental reflexive (post-)modernity, a simple before and after' (Swyngedouw and Ernstson 2018: 9). In such a representation, science is typically presented as core to the undesirable modernist nature-society split that a more reflexive Anthropocene worldview needs to overcome. Without negating the importance of the latter, such a representation of science denies the key ongoing role that certain sciences have played in enabling the sort of informed reflexivity needed (see Beck 1992). It thus overlooks the fact that 'Modernity is not a single-headed process that now has been surpassed. On the contrary; it is the multi-headed internal struggle that predates as well as postdates the event of the Anthropocene' (Swyngedouw and Ernstson 2018: 9).

Second, and relatedly, black-boxing science brackets out the powerful political economic context of specific scientific groups and endeavours, contributing to the normalisation of science's existing, often involuntary, weaponisation in many efforts to legitimise actions that simply worsen Anthropocene conditions.[11] While vital issues are exposed by contrasting a monolithic 'Western science' with other ontologies and cosmologies (e.g. Simpson 2018), there is a parallel need for more nuanced engagement with the varied content, sites and use of science. In particular, there is a need to engage with those fighting to ensure the integrity of science and the public good of its outcomes – not in terms of lofty declarations about science in society which equally are guilty of black-boxing, but in terms of real tussles over the use of science in the world. In this way, there is, as Castree (2015) suggests, an important opportunity for critical social scientists to work with willing geoscientists to (re)politicise their work towards progressive ends. Engaging with existing work by groups such as the Concerned Scientists above, and/or welcoming scientists into wider alliances fighting for similar ends (e.g. the Stop Adani social movement) promises to broaden the front on which the fight against the worsening of Anthropocene conditions can be fought and to build vital new alliances across reified boundaries. This is not just about enrolling scientists as activists (e.g. in civil disobedience against the Keystone XL Pipeline, see Bradshaw 2015) but about helping them (re)weaponise their scientific practices and findings, and recognising that not only natures but science and scientists act in unpredictable ways.

In sum, the ongoing acceleration of Anthropocene conditions requires that better science (that is more accurate, multi-scalar, precautionary and 'objective' in a Haraway sense) contests the weaponisation of other (pseudo)science to legitimise destructive activities such as the Adani mine. Given this situation, it is unhelpfully ironic if social scientists dismiss all science out of hand as part of the problem. Just as critical social science offers a weapon of improved contextual understanding to scientists, strong, accurate and critical science offers a weapon for progressive politics.

Towards a 'world ironic relation'?

For those seeking to stem the Earth's descent into an Anthropocene state, the ironies demanding attention do not end with communicative or even situational ironies. Instead, Szerszynski (2007) suggests that we need to try to stay ahead of the ongoing generation and layering of contradictions by moving beyond specific, disappointing ironies to embracing irony in general at a higher level. Irony emerges here as actually a form of freedom. Szerszynski echoes Foucault in drawing on Danish existentialist philosopher Søren Kierkegaard, author of *The Concept of Irony* (1971), to examine how people cope with the disorientation and confusion resulting from the fundamental absurdities of life (Flynn 2006; Flynn 1997). Kierkegaard argued for an ironic stance, one directed 'not against this or that particular existence but against the whole given actuality of a certain time and place' (Kierkegaard 1971 in Donoghue 2014: 138). Rather than retreating from irony or countering its apathetic tendencies with a zealous determination to eliminate inconsistencies, this suggests we need to go deeper into the ironies of our entire situation while not abandoning our effort to enact real world change. Drawing on Mueke (1969), Szerszynski terms this approach an 'ironic world relation': 'a generalised ironic stance towards the world and oneself'. Rather than adopting the role of a 'distanced observer, aloof from the folly and blindness they perceive being played out in front of them', a metaphysical ironic world relation 'embraces even the observer, the identifier of the irony, within its grasp' (Szerszynski 2007: 348).

Such a stance shares with both the turn to resilience thinking and adaptive management, and more-than-humanist ontologies, a common interest in the agency of more-than-humans (see Braun 2015a). However, it expands their interest in agency and indeterminacy to life and existence in general. In doing so, it echoes Donna Haraway (1988) who noted that:

> Acknowledging the agency of the world in knowledge makes room for some unsettling possibilities, including a sense of the world's independent sense of humor. Such a sense of humor is not comfortable for humanists and others committed to the world as resource.
>
> *(Haraway 1988: 593)*

By expanding an appreciation of agency to the world in general, an ironic 'world relation' encourages a search for the unexpected in the seemingly familiar, including knowledges and groups. For example, while science and scientists are frequently viewed (negatively or instrumentally) as a resource, they are also, as we have seen, capable of surprise and humour including activism, though of a less familiar legal and scientific form than picket line protests. An ironic relation tries to escape the categorisation of this or that point of view or action according to some stable notion of its starting point or identity, and instead emphasises the uncertainty of such categories. Claire Colebrook (2000) argues that in modern irony, communicative irony is undermined by the fact that 'it is precisely attribution, position or point of view which is rendered problematic' (2000: 16). Among what is rendered problematic are, as Szerszynski notes, our own identity and position. Who and where we are and what we mean or what we can do is never as mappable, stable or knowable as we may assume. As Colebrook continues:

when we look at what we say we no longer understand what we mean [...]
The performance of irony demonstrates that what we say might always be
more than what we assume or what we recognize as being our own, our
context.

(Colebrook 2000: 18)

It is thus not only the future that we do not know in the Anthropocene, it is the
present and the past; it is not only the planet that is rendered unfamiliar, it is our
own voice and actions.

In such a situation – faced with the need to do something but not knowing
where to start – it makes sense to go back to basics: to go back to the context each
of us finds ourselves in and identify what is happening. To do this we need, as
Haraway notes, to combine a commitment to the situatedness of knowledge
(broadly defined) with 'a no-nonsense commitment to faithful accounts of a "real"
world' (Haraway 1988: 579). For a growing number of scholars, this is less about
honing our established critiques than honing in on where the politics lies – in the
sense of Ranciere's (1998) idea of politics as 'the disruption of the sensible'. Rather
than just expecting or striving to help the existing order recover (i.e. to be resilient), there is a growing need for political practices that maintain 'fidelity to the
truth of the interrupting event' such that we move and think with the situation, in
accordance with the event (*cf* Swyngedouw and Ernstson 2018 on Badiou 2012).
Calls to hone in on where disruptions to the sensible are evident resonate strongly
with the primary message of this chapter: to be alert to the ironies that we are
entwined with and (counter to what Swyngedouw and Ernstson 2018 conclude)
this chapter has also indicated that it is possible and useful to understand the discursive event of the Anthropocene in this way.

Although everything in the Anthropocene has been rendered 'experimental [and
ironic] top to bottom' (Braun 2015b: 103), and our firm conceptual foundations
seem to have been lost, the resultant confusion and academic in-fighting does not
remove the need to address the proliferating empirical situations where 'sides have
to be taken, choices made' (Swyngedouw and Ernstson 2018: 21). Ironic ecology
demands 'recognition of failure and error' at the same time as 'the need to act, with
due care, in the very face of that recognition' (Szerszynski 2007: 351). Given such a
situation, the appearance of ironies (such as the abuse of the concept of adaptive
management in the mining industry) can act as a signpost and motor for critique,
not a smokescreen or brake. This may be why issues such as the Adani mine and
Keystone XL pipeline attract such a broad range of support – because in the face of
even just the possibility that the planet is entering an Anthropocene state, preventing further massive fossil fuel developments is clearly necessary. At the same
time, the ironies that each of these cases involve (thanks in part to their entanglement with post-truth politics) means that there is a ready list of tasks for critical
scholars to get stuck into. When the wider ironies of the sort outlined in this
chapter are taken into account, further tasks are also exposed. This includes the
need to engage empathetically with the efforts of dissenting scientists whose starting

point is often the cynical abuse of their tools and concepts, as well as the need to engage with satire and humour, not the least to maintain our mental resilience. It may be that even if we adopt an 'ironic world relation' in the end we are 'hood-winked' by the world (*cf* Haraway 1988) just as we always are; but in the face of mounting ironies and eroding foundations, we may as well jump in and see if it helps us overcome the ecological paradox of informed inaction.

Notes

1 See Merchant (2003) and Vann (2012) on these powerful grand narratives in the Abra-hamic cultural traditions.
2 http://throbgoblins.blogspot.com.au/
3 See http://throbgoblins.blogspot.com.au/search?q=frank
4 For more of Marc Roberts' cartoons, see https://marcrobertsink.wordpress.com/art/
5 https://theconversation.com/adani-is-cleared-to-start-digging-its-coal-mine-six-key-questions-answered-118760
6 https://www.edgehill.ac.uk/news/story/habitus-art-installation-unveiled/
7 See https://www.economist.com/leaders/2011/05/26/welcome-to-the-anthropocene and http://www.futureearth.org/asiacentre/welcome-anthropocene
8 http://throbgoblins.blogspot.com.au/2011/08/all-gone-very-quiet-hasnt-it.html Acces-sed Nov 15, 2014
9 See also https://www.eurekalert.org/pub_releases/2019-06/fu-lat061019.php
10 See for example https://www.theguardian.com/australia-news/2019/feb/20/murray-da rling-authority-promises-to-upgrade-climate-science-after-criticism See also Alexandra (2018; 2019)
11 On the conservative movement's use of 'science' to support the fossil fuel industry in the US see McCright and Dunlap (2010).

References

Alexandra, J. (2018) Evolving governance and contested water reforms in Australia's Murray Darling Basin. *Water* 10: 113–131.
Alexandra, J. (2019) Losing the authority – what institutional architecture for cooperative governance in the Murray Darling Basin? *Australasian Journal of Water Resources* 23(2): 99–115.
Angus, I. (2016) *Facing the Anthropocene: Fossil Capitalism and the Crisis of the Earth System.* New York: NYU Press.
Autin, W. J. and Holbrook, J.M. (2012) Is the Anthropocene an issue of stratigraphy or pop culture? *GSA Today* 22: 60–61.
Badiou, A. (2012) *Ethics: An Essay on the Understanding of Evil.* London: Verso.
Bailey, I., Gouldson, A. and Newell, P. (2011) Ecological modernisation and the governance of carbon: a critical analysis. *Antipode* 43: 682–703.
Beck, U. (1992) *Risk Society: Towards a New Modernity.* London: Sage.
Benson, M. H. and Garmestani, A. S. (2011) Embracing panarchy, building resilience and integrating adaptive management through a rebirth of the National Environmental Policy Act. *Journal of Environmental Management* 92: 1420–1427.
Blühdorn, I. (2007) Sustaining the unsustainable: symbolic politics and the politics of simu-lation. *Environmental Politics* 16: 251–275.
Blühdorn, I. (2011) The politics of unsustainability: COP15, post-ecologism, and the eco-logical paradox. *Organization & Environment* 24: 34–53.

Boykoff, M. and Osnes, B. (2019) A laughing matter? Confronting climate change through humor. *Political Geography* 68: 154–163.

Bradshaw, E. A. (2015) Blockadia rising: rowdy greens, direct action and the Keystone XL pipeline. *Critical Criminology* 23: 433–448.

Braun, B. (2015a) The 2013 Antipode RGS-IBG Lecture – New Materialisms and Neo-liberal Natures. *Antipode* 47: 1–14.

Braun, B. (2015b) From critique to experiment? Rethinking political ecology for the Anthropocene. In T. Perreault, G. Bridge and J. Mccarthy (eds), *The Routledge Handbook of Political Ecology*. London: Routledge, 102–114.

Carpenter, S., Walker, B., Anderies, J. M. and Abel, N. (2001) From metaphor to measurement: resilience of what to what? *Ecosystems* 4: 765–781.

Carty, J. and Musharbash, Y. (2008) 'You've got to be joking': asserting the analytical value of humour and laughter in contemporary anthropology. *Anthropological Forum* 18: 209–217.

Castree, N. (2014) Geography and the Anthropocene II: current contributions. *Geography Compass* 8: 450–463.

Castree, N. (2015) Unfree radicals: geoscientists, the Anthropocene, and left politics. *Antipode* 49: 52–74.

Clémençon, R. (2012) Welcome to the Anthropocene: Rio+20 and the meaning of sustainable development. *The Journal of Environment & Development* 21: 311–338.

Colebrook, C. (2000) The meaning of irony. *Textual Practice* 14: 5–30.

Craig, R. K. and Ruhl, J. (2014) Designing administrative law for adaptive management. *Vanderbilt Law Review* 67: 1–88.

Crist, E. (2013) On the poverty of our nomenclature. *Environmental Humanities* 3: 129–147.

Currell, M. J., Werner, A. D., Mcgrath, C., Webb, J. A. and Berkman, M. (2017) Problems with the application of hydrogeological science to regulation of Australian mining projects: Carmichael Mine and Doongmabulla Springs. *Journal of Hydrology* 548: 674–682.

DiCaglio, J. (2014) Ironic ecology. *Interdisciplinary Studies in Literature and Environment* 22: 447–465.

Dittmer, J. (2005) Captain America's empire: reflections on identity, popular culture, and post-9/11 geopolitics. *Annals of the Association of American Geographers* 95: 626–643.

Donoghue, D. (2014) *Metaphor*. Cambridge, MA: Harvard University Press.

Eriksen, C. (2019) Negotiating adversity with humour: a case study of wildland firefighter women. *Political Geography* 68: 139–145.

Fisher, D. E. (2014) The rule of law, the public interest and the management of natural resources in Australia. *Environmental and Planning Law Journal* 31: 151–163.

Fluri, J. L. and Clark, J. H. (2019) Political geographies of humor and adversity. *Political Geography* 68: 122–124.

Flynn, T. R. (1997) 'Sartre, Foucault and Historical reason, vol. 1: toward an existentialist theory of history. *Revue Philosophique de la France Et de l'Etranger* 188: 498–500.

Flynn, T. (2006) *Existentialism*. Oxford: Oxford University Press.

Folke, C., Hahn, T., Olsson, P. and Norberg, J. (2005) Adaptive governance of social-eco-logical systems. *Annual Review of Environment and Resources* 30: 441–473.

Gaud, W. S., Price, P. W. and Slobodchikoff, C. (1984) *A New Ecology: Novel Approaches to Interactive Systems*. London: Wiley.

Gibbs, R. W. (2002) Irony in the wake of tragedy. *Metaphor and Symbol* 17: 145–153.

Grove, K. (2017) *Resilience*. Abingdon: Routledge.

Hamilton, C. (2016) Define the Anthropocene in terms of the whole Earth. *Nature* 536: 251.

Haraway, D. (1988) Situated knowledges: the science question in feminism and the privilege of partial perspective. *Feminist Studies* 14: 575–599.

Hoggett, P. and Randall, R. (2018) Engaging with climate change: comparing the cultures of science and activism. *Environmental Values* 27: 223–243.

Holling, C. S. (1978) *Adaptive Environmental Assessment and Management*. London: John Wiley & Sons.

Holling, C. S. and Meffe, G. K. (1996) Command and control and the pathology of natural resource management. *Conservation Biology* 10: 328–337.

Hutcheon, L. (1994) *Irony's Edge: The Theory and Politics of Irony*. London: Routledge.

Karlsson, R. (2013) Ambivalence, irony, and democracy in the Anthropocene. *Futures* 46: 1–9.

Kelly, J. M., Scarpino, P., Berry, H., Syvitski, J. and Meybeck, M. (2018) *Rivers of the Anthropocene*. Berkeley: University of California Press.

Kierkegaard, S. (1971) *Either/Or*. Translated by D.F. Swenson. Princeton NJ: Princeton University Press.

Kuus, M. (2008) Švejkian geopolitics: subversive obedience in Central Europe. *Geopolitics* 13: 257–277.

Lee, J. (2014) Theory to practice: adaptive management of the groundwater impacts of Australian mining projects. *Environmental and Planning Law Journal* 31: 251–287.

Lekan, T. M. (2014) 'Fractal Earth: Visualizing the Global Environment in the Anthropocene', *Environmental Humanities* 5: 171–201.

Lewandowsky, S., Oberauer, K. and Gignac, G. E. (2013) NASA faked the moon landing – therefore, (climate) science is a hoax: an anatomy of the motivated rejection of science. *Psychological Science* 24: 622–633.

Lewandowsky, S., Oreskes, N., Risbey, J. S., Newell, B. R. and Smithson, M. (2015) Seepage: climate change denial and its effect on the scientific community. *Global Environmental Change* 33: 1–13.

Lilley, S., Mcnally, D., Yuen, E. and Davis, J. (2012) *Catastrophism: The Apocalyptic Politics of Collapse and Rebirth*. Oakland, CA: PM Press.

Lövbrand, E., Beck, S., Chilvers, J., Forsyth, T., Hedrén, J., Hulme, M., Lidskog, R. and Vasileiadou, E. (2015) Who speaks for the future of Earth? How critical social science can extend the conversation on the Anthropocene. *Global Environmental Change* 32: 211–218.

Lynas, M. (2011) *The God Species: Saving the Planet in the Age of Humans*. Boone, IA: National Geographic Books.

Maibach, E., Myers, T. and Leiserowitz, A. (2014) Climate scientists need to set the record straight: There is a scientific consensus that human-caused climate change is happening. *Earth's Future* 2: 295–298.

Manzo, K. (2012) Earthworks: the geopolitical visions of climate change cartoons. *Political Geography* 31: 481–494.

McCright, A. M. and Dunlap, R. E. (2010) Anti-reflexivity: the American conservative movement's success in undermining climate science and policy. *Theory, Culture & Society* 27: 100–133.

McCullough, M. B. (2008) 'Poor black bastard can't shake-a-leg': humour and laughter in urban Aboriginal North Queensland, Australia. *Anthropological Forum* 18: 279–285.

McGinley, K. (2017) Adapting tropical forest policy and practice in the context of the Anthropocene: opportunities and challenges for the El Yunque National Forest in Puerto Rico. *Forests* 8: 259.

Merchant, C. (2003) *Reinventing Eden: The Fate of Nature in Western Culture*. New York: Routledge.

Mirowski, P. (2019) Hell is truth seen too late. *boundary 2: an international journal of literature and culture* 46: 1–53.

Monnat, S. M. and Brown, D. L. (2017) More than a rural revolt: landscapes of despair and the 2016 Presidential election. *Journal of Rural Studies* 55: 227–236.

Morton, T. (2007) *Ecology Without Nature: Rethinking Environmental Aesthetics*. Cambridge, MA: Harvard University Press.

Morton, T. (2010) *The Ecological Thought*. Cambridge, MA: Harvard University Press.

Muecke, D. C. (1969) *The Compass of Irony*. London: Methuen.

Neimark, B., Childs, J., Nightingale, A. J., Cavanagh, C. J., Sullivan, S., Benjaminsen, T. A., Batterbury, S., Koot, S. and Harcourt, W. (2019) Speaking power to 'post-truth': critical political ecology and the new authoritarianism. *Annals of the American Association of Geographers* 109(2): 612–623.

Nullmeier, F. (2005) Nachwort. In M. Edelman (Ed), *Politik als Ritual. Die Symbolische Funktion staatlicher Institutionen und politischen Handelns*, 3rd edn, pp. 199–219. Frankfurt: Campus.

Oxford English Dictionary (2015) 3rd edition. Edited by A. Stevenson. Oxford: Oxford University Press.

Pellizzoni, L. (2014) Risk. In C. Death (ed.), *Critical Environmental Politics: Interventions*. Abingdon: Routledge, 198–207.

Pellizzoni, L. (2016) *Ontological Politics in a Disposable World: The New Mastery of Nature*. Abingdon: Routledge.

Phillips, D. (2015) Posthumanism, environmental history, and narratives of collapse. *Interdisciplinary Studies in Literature and Environment* 22: 62–79.

Pickering, J. (2018) Ecological reflexivity: characterising an elusive virtue for governance in the Anthropocene. *Environmental Politics* 28(7): 1145–1166.

Ranciere, J. (1998) *Disagreement*. Minneapolis: University of Minnesota Press.

Rickards, L. (2015) Metaphor and the Anthropocene: presenting humans as a geological force. *Geographical Research* 53: 280–287.

Ridanpää, J. (2009) Geopolitics of humour: the Muhammed cartoon crisis and the Kaltio comic strip episode in Finland. *Geopolitics* 14: 729–749.

Ridanpää, J. (2010) A masculinist northern wilderness and the emancipatory potential of literary irony. *Gender, Place & Culture: A Journal of Feminist Geography* 17: 319–335.

Ridanpää, J. (2014) Geographical studies of humor. *Geography Compass* 8: 701–709.

Ridanpää, J. (2019) Crisis events and the inter-scalar politics of humor. *GeoJournal* 84: 901–915.

Robson, M. (2019) Metaphor and irony in the constitution of UK borders: An assessment of the 'Mac' cartoons in the Daily Mail newspaper. *Political Geography* 71: 115–125.

Ronan, M. (2017) Religion and the environment: twenty-first century American evangelicalism and the Anthropocene. *Humanities* 6: 1–15.

Rull, V. (2013) A futurist perspective on the Anthropocene. *The Holocene* 23: 1198–1201.

Sayre, N. F. (2012) The politics of the anthropogenic. *Annual Review of Anthropology* 41: 57–70.

Schnaiberg, A. (1980) *The Environment: From Surplus to Scarcity*. New York: Oxford University Press.

Scoones, I., Edelman, M., BorrasJr, S. M., Hall, R., Wolford, W. and White, B. (2018) Emancipatory rural politics: confronting authoritarian populism. *The Journal of Peasant Studies* 45: 1–20.

Shellenberger, M. and Nordhaus, T. (2004) *The Death of Environmentalism*. Breakthrough Institute. https://thebreakthrough.org/articles/the_death_of_environmentalism

Shewry, T. (2019) 'Ice thieves: urban water, climate justice, and the humor of incongruity' in Jane Rawson's A Wrong Turn at the Office of Unmade Lists. *English Language Notes* 57: 82–95.

Simpson, M. (2018) The Anthropocene as colonial discourse. *Environment and Planning D: Society and Space*, online first, 1–19.

Steffen, W., Grinevald, J., Crutzen, P. and McNeill, J. (2011) The Anthropocene: conceptual and historical perspectives. *Philosophical Transactions of the Royal Society A: Mathematical, Physical and Engineering Sciences* 369: 842–867.

Steffen, W., Leinfelder, R., Zalasiewicz, J., Waters, C., Williams, M., Summerhayes, C., ... Schellnhuber, H. (2016) Stratigraphic and Earth System approaches to defining the Anthropocene. *Earth's Future* 4: 324–345.

Steffen, W., Rockström, J., Richardson, K., Lenton, T. M., Folke, C., Liverman, D., ... Schellnhuber, H. J. (2018) Trajectories of the Earth System in the Anthropocene. *Proceedings of the National Academy of Sciences* 115: 8252–8259.

Swan, E. and Fox, S. (2009) Becoming flexible: self-flexibility and its pedagogies. *British Journal of Management* 20: 149–159.

Swyngedouw, E. (2007) Impossible 'sustainability' and the post-political condition. In R. Krueger and D. Gibbs (eds), *The Sustainable Development Paradox*. New York: Guilford Press, 13–40.

Swyngedouw, E. (2010) Apocalypse forever? Post-political populism and the spectre of climate change. *Theory, Culture & Society* 27: 213–232.

Swyngedouw, E. (2013) Apocalypse now! Fear and Doomsday pleasures. *Capitalism Nature Socialism* 24: 9–18.

Swyngedouw, E. and Ernstson, H. (2018) Interrupting the Anthropo-obScene: immunobiopolitics and depoliticizing more-than-human ontologies in the Anthropocene. *Theory, Culture & Society* 35: 3–30.

Szerszynski, B. (2007) The post-ecologist condition: irony as symptom and cure. *Environmental Politics* 16: 337–355.

Tadaki, M., Brierley, G., Dickson, M., Le Heron, R. and Salmond, J. (2014) Cultivating critical practices in physical geography. *The Geographical Journal* 181: 160–171.

Vann, B. A. (2012) *Forces of Nature: Our Quest to Conquer the Planet*. New York: Prometheus Books.

Volkmar, A. (2017) Ironic encounters in the Anthropocene: Jürgen Nefzger's nuclear landscape photography. *Frame. Tijdschrift voor Literatuurwetenschap* 29: 46–69.

Walker, J. and Cooper, M. (2011) Genealogies of resilience: from systems ecology to the political economy of crisis adaptation. *Security Dialogue* 42: 143–160.

Werner, A. D., Love, A., Irvine, D., Banks, E., Cartwright, I., Webb, J. and Currell, M. (2019) Position paper by concerned scientists: deficiencies in the scientific assessment of the Carmichael Mine impacts to the Doongmabulla Springs. https://dspace.flinders.edu.au/xmlui/handle/2328/39203

Whitehead, M. (2014) *Environmental Transformations: A Geography of the Anthropocene*. London: Routledge.

Zimmerer, K. S. (1994) Human geography and the 'new ecology': the prospect and promise of integration. *Annals of the Association of American Geographers* 84: 108–125.

9

'PRIMORDIAL WOUNDS'

Resilience, trauma and the rifted body of the Earth

Nigel Clark

Introduction: fantastic voyages

The widening rift between the two superpowers in the post-war decades gave a boost to Earth science research. Cold War projects generated novel understandings about the integration of the atmosphere and hydrosphere, advanced the study of the planet-wide cycling of the major chemical elements, and provided the evidence of sea-floor spreading at mid-ocean ridges that helped confirm the theory of plate tectonics (Masco 2010; Davis 1996). Laypeople, in the west at least, may have struggled to get their heads around the global scale of these 'geophysical' and 'geopolitical' forces. But Hollywood stepped up to assist. The 1966 science fiction film *Fantastic Voyage* rescaled superpower rivalry and the adventure of the abysmal depths to a decidedly intimate level: the now iconic storyline involving a deep ocean submarine miniaturised to make life-saving medical interventions within the body of a scientist who has suffered brain injury in the course of a Cold War assassination attempt.

Fast-forward 50 years and another, less fictional, fantastic journey is underway. Marine scientists, it was announced in 2007, were venturing into the deep Atlantic – using sonar and robotic seabed drills seabed to investigate what they described as an 'open wound' in the planetary body (Than 2007). As the submarine geology of the late 1950s and early 1960s revealed, new crust is formed as magma rising up from the inner Earth pumps out along mid-oceanic plate junctures – where it quickly hardens into basalt rock. But researchers have identified an anomaly on the floor of the mid-Atlantic. Here, in a zone covering thousands of square kilometres, rocky crust has failed to form – and the Earth's interior mantle layer remains exposed. As a science reporter explains, citing one of the marine geologists taking part in the expedition: 'MacLeod likens this process to stretching a person's skin until it ruptures, exposing the flesh underneath. "You take the crust and you stretch it and you pull it and pull it until it breaks," he said' (Than 2007).

Natural scientists often deploy tropes of cuts, gouges, gashes and other bodily injury in this way to help audiences empathize with the inhuman vastness of violent Earth processes. Social thinkers, on the other hand, have long borrowed geologic imagery – rifts, ruptures, chasms, seismic shifts – to portray events so momentous that they affect our very ability to make sense of the world. In the context of the Anthropocene – with its scale-bending narrative of diminutive beings inflicting grievous injury on an astronomical body – it is not always clear which direction such traffic in signifiers is travelling. Ethical philosopher Clive Hamilton and science historian Jacques Grinevald describe the Anthropocene as 'a new anthropogenic rift in the natural history of planet Earth' (Hamilton and Grinevald 2015: 9). Extrapolating on Marx's proto-ecological understanding of metabolic rifts between city and countryside, Marxist political ecologists John Bellamy Foster, Brett Clark and Richard York diagnose 'an extreme "rift" in the planetary system' (Foster et al. 2010: 15). With the coming of the climate change and Anthropocene, literary theorist Tim Morton contends, the geosciences confront 'an abyss whose reality becomes increasingly uncanny, not less, the more scientific instruments are able to probe it' (Morton 2012: 233).

How we approach the concept of resilience, I will be suggesting, is closely associated with our imagining of injury, vulnerability and trauma. But in turn, how we think about our own exposure – our fleshy and psychic susceptibility – is bound up with the different ways in which we conceive of the rifting and rumbling of our geophysical environment. In the midst of the Anthropocene event this exchange between tropes of planetary and bodily perishability is not only more complex than ever, I contend, but embodies a fundamental tension. Is harm largely generated *within* a body or system – or does it come from a potentially hostile *exterior*? To put it another way, is the pursuit of resilience a question of responding to the threats of self-inflicted injury – or is it more about coming to terms with a basic perviousness and exposure to outside forces? The weighting we put on these two options matters, I argue, and it is closely tied to differences within the geoscience community who are exploring the Anthropocene thesis.

Addressing possible precursors to Anthropocene thesis, Hamilton and Grinevald (2015) are insistent that only the newish interdisciplinary field of Earth system science can provide the requisite understanding of the human capacity to disrupt the great flows and cycles that compose our planet. Through its definitive concern with global biogeochemical cycles and their interactions, Earth system science demonstrates how it is possible for the planet to generate its own transformations at every scale – from the localized ecosystem right up to the total planetary system. As leading Earth system scientist Will Steffen puts it: 'the forcings and feedbacks within the Earth System are as important as the external drivers of change, such as variability in solar energy input' (Steffen and Lambin 2006: 1). In this context, the thematic of resilience is primarily focused on the capacity of any system to resist pressure for change – and in this regard is considered the 'antonym' of vulnerability (Folke in Steffen et al. 2004: 287; Steffen et al. 2004: 205). Whereas in a vulnerable system even small changes may nudge the system over a threshold into a new operating state, a resilient system is one with the capacity to maintain its existing state by absorbing stresses or bouncing backing from perturbations (Folke in Steffen et al. 2004: 287).

While a formative influence on Earth system science was the idea that life itself has played a key role in planetary dynamics for billions of years – as proposed in the Gaia hypothesis (Steffen et al. 2004: 3), it is important to keep in mind that the field came of age focusing on 'contemporary global change' – which is to say, change primarily induced by our own species (Zalasiewicz et al. 2017: 3). However, when it comes to gathering evidence to formalize the Anthropocene as a geological epoch, the responsibility falls largely on the older field of stratigraphic geology – whose definitive concern, as Anthropocene Working Group chair Jan Zalasiewicz and his colleagues sum up, is 'with ancient, pre-human rock and time' (Zalasiewicz et al. 2017: 3). Perhaps the signal achievement of Anthropocene science thus far has been the way that it has brought these two formerly distinct geoscience fields together – in particular to collaborate over the question of what traces human-triggered transformation in the present Earth system will likely leave in the lithic strata of the distant future (see Zalasiewicz et al. 2017; Steffen et al. 2016).

Promising though this alliance is, we need to be as attentive to the differences between stratigraphic geology and Earth system science as we are to their convergence. And to attend to this distinction, I would add, for reasons other than those most critical social scientists have thus far alighted upon when they have taken Anthropocene science to task. Earth system science has frequently come under fire for presenting a unified, undifferentiated figure of 'the human' and for insufficiently specifying the socio-economic processes that underpin planetary change (see Clark and Gunaratnam 2017). However, the interdisciplinary field's axiom that 'social-ecological systems act as strongly coupled, integrated complex systems' has been more broadly accepted by social scientists, most likely because of the way it endorses and extends the sphere of influence of social thought (Folke 2004: 287, see also Bai et al. 2016; Lovbrand et al. 2015). Conversely, with its foregrounding of great temporal and spatial reaches where no human presence is to be found, stratigraphic geology is more likely to be targeted by critical thinkers for its exteriorization of nature – and the unreconstructed duality of subject and object this is said to imply (see Clark and Gunaratnam 2017).

It is my aim here to invert – or pervert – this logic. Looking more closely at Earth system science's investment in the co-extensiveness and mutuality of human and biogeophysical systems, I ask what is excluded by the depreciation of radical exteriority – and what the implications of this might be for understanding resilience. Turning to 'older school' geology, I pick up on its thematizing of a planetary surface that remains constitutively – and asymmetrically – open to the forces of the inner Earth. It is this holding open of the space-times of life to the unliveable, unassimilable powers of the planetary body, I propose, that can and should implicate geological imaginaries in a vision of resilience that is less sharply distinguished from or opposed to vulnerability. Drawing on Sigmund Freud's prefigurative linking of psychic and somatic trauma, and its wild extrapolation by philosophers

Nick Land and Reza Negarestani, I gravitate towards a notion of resilience that cannot be separated from an inaugurating and ongoing 'geotraumatic' exposure to the violence of the Earth and cosmos. More than a matter of negotiating between different scientific framings of the vulnerability-resilience nexus, I suggest, this has profound implications for the way critical social scientists might view the relationship between human collectives and our geocosmic abode.

Closure and opening in the Earth system

The self-understanding of Earth system science is of a bold embrace of complexity and dynamism in the physical world. Turning away from the gradualism that reigned in the Earth and life sciences since the early 19th century, and moving beyond the ideal of equilibrium or steady states that prevailed in mid 20th century systems thinking, the interdisciplinary geoscience that came of age in the 1980s coalesced around the idea that 'the Earth is never static and … variability abounds at nearly all spatial and temporal scales' (Steffen et al. 2004: 295). The key to this sense of constant potential for change lies in the way that dense interconnections within complex systems organize themselves into feedback loops. Feedbacks – which involve the recursive cycling of inputs or effects through a system – can serve either to absorb and dampen down pressures for change or to amplify them: they can be 'healing' or 'hurting' as Anthropocene progenitor Paul Crutzen (2004: 72) puts it. By this logic, complex systems are in a constant process of responsiveness and self-adjustment – but if pushed beyond a certain threshold they have the capacity to shift rapidly from one overall operating state to another.

It is this thematizing of inherent changeability that informs Earth system science's preference for notions of resilience over sustainability. Whereas sustainability is seen to rest on assumptions that at any moment we can identify what it is we are trying to protect, the concept of resilience acknowledges the uncertainty that is the essential correlate of the system's constant flux and ongoing transmutability (Folke in Steffen et al. 2004: 287; Benson and Craig 2014). The challenge then is not only a matter of how to reduce the pressures that might nudge a system out of a 'desirable state', but how to protect and reinforce the system's own capacity to adapt to stress.

What the Anthropocene thesis brings into relief is that these imperatives apply at the scale of the total Earth system. But this also has profound implications for how 'we' conceive of ourselves. As a source of 'stressors' or pressures to change, humankind is conceived as part of the overall system: an increasingly forceful subcomponent *within* dynamic interactions of the various cycles and fluxes of the Earth system. In this light, the onus is not only on us to modulate our impacts on the various components of the Earth system – but also to cultivate our own adaptive capacities in the face of an increasingly uncertain global environment.

It is the apparent resonance between this 'scientific' imperative to enhance human resilience and the demands foisted on disaggregated social actors by the current phase of the global capitalism that has aggravated critical social thinkers. I want to take a different approach, however, which is to ask what is at stake in binding social systems so tightly into the systemicity of the Earth that the human and the nonhuman are rendered co-extensive. Such a concern means that rather than pushing forward and zooming in on the internal differentiation of the 'Anthropos', we step back and consider how Earth system scientists define the object of their inquiry – prompting us to ask what is included in the Earth system, what is excluded, and what the implications of these 'cuts' might be.

It is telling that when it comes to one of the most basic categorizations of a system – is it open or closed? – it is difficult to get a clear-cut answer about our planet from Earth systems scientists. The formal definition of closed system that it is open to energy flows but has no exchange of matter, by which logic the Earth – powered by the sun but with limited extra-planetary transfer of matter – largely fits the description. Such a classification justifies prioritization of the interacting circulations, fluxes and reservoirs *within* the Earth system. As Earth system scientist Tim Lenton puts it in *Earth System Science: A Very Short Introduction*: '[i]t is the thin layer of a system at the surface of the Earth – and its remarkable properties – that is the subject of my work' (Lenton 2016: 17).

Philosopher and science studies scholar Bruno Latour cites this passage from Lenton approvingly (Latour 2014: 8). Latour likewise announces his focus on the 'envelope of the biosphere' – which he follows up through his engagement in the study of 'critical zones', a way of applying Earth system science to vertical slices of that section of the Earth where living things interact with the inorganic world. As he explains: 'critical zones define a set of interconnected entities in which the human multiform actions are everywhere intertwined' (Latour 2014: 3).

At this point it's worth recalling Earth system science's formative interest in human-induced global change. For there is a question here of whether prevailing constructions of the Earth system reflect an 'impartial' appraisal of planetary functionality or whether the constitutive concern with human-nonhuman coupling is itself a marker of what counts as systemicity. To put it another way, in the process of reimagining the social as one more component of the Earth system, might a certain cutting of the Earth processes to fit the measure human be taking place? For if we look more closely at what precedes Lenton's selection of that section of the planet that happens to support life, the self-evidence of the definitive cut is rather less obvious. As Lenton elaborates:

> What is less clear is whether and where to put an inner boundary on the Earth system ... The longer the timescale we look over, the more we need to include in the Earth system ... material in the Earth's crust becomes part of the Earth system, and we must recognize that the crust also exchanges material with the Earth's mantle.
>
> *(Lenton 2016: 16)*

As he goes on to say: 'For many Earth system scientists, the planet Earth is really comprised of two systems – the surface Earth system that supports life, and the great bulk of the inner Earth underneath' (Lenton 2016: 17). And this implies that at some point the logic of coupling – or Latour's 'interwining' or 'interconnection' – reaches a limit. As a current hypothesis would have it, the lithosphere – the relatively rigid outer layer of the Earth – is *decoupled* from the underlying Aesthenosphere – the upper layer of the Earth's mantle in which hot viscous rock slowly cycles in vast convection currents (Self and Rampino 2012). But 'decoupling' here doesn't imply separation. What it means is that there an asymmetrical relationship, a grounding or subtending relation rather than the tightly configured interdependence that characterizes the 'envelope' of biogeochemical cycles which is of most interest to Lenton and the majority of his Earth system science colleagues.

For the older school 'pre-human rock and time' geologists, this underpinning of the crustal structure by the forces of the inner Earth is pivotal for understanding how rocky strata are formed, deformed and rearranged. As Zalasiewicz points out, compared with other astronomical bodies in this solar system, our planet has extraordinarily rich and diverse lithic strata (Zalasiewicz 2008: 17). While the constantly moving envelope of gaseous, liquid and biotic processes on and around the planet's surface play a crucial role in stratal formation, it is the Earth's exceptionally mobile plate tectonics that propel this process. In turn, it is the convection of the mantle layer that drives tectonic plate motion – all ultimately powered by heat dissipating upwards from the planet's core (Zalasiewicz 2008: 46–51).

Mantle activity, in this way, constructs the sea floor, shifts continents and oceans around, builds mountains and valleys, and generates new lithic strata. As paleontologist Richard Fortey would have it, the power of the inner Earth 'is the deep unconscious of our planet, the hidden body whose bidding the continents obey' (Fortey 2005: 414). It is not only a matter of motive force, of driving the construction of planetary topography, however. The ongoing upward surges of mantle rock rendered molten in the course of tectonic plate movement also provide many of the materials upon which biological life depends. '[F]rom the secret storehouses of the inner earth,' observes science writer Simon Winchester, come 'the elements that allow the outer earth, the biosphere, the lithosphere, to be so vibrantly alive' (Winchester 2004: 302).

To the idea that the life-sustaining envelope of the outer Earth is open to the forces of the inner Earth we must add the thesis that inbound astronomical bodies – meteor impacts – played a formative role in the making of the planet's crust and periodically add their mighty perturbations to the trajectory of life (see Davis 1996; Brooke 2014: 26–37). Such a vision of combined endogenous and exogenous planetary 'forcings', however, has very different implications from the assumption that most of the significant planetary action takes place in the slender inter-zone where humans and other life forms are entangled with inorganic processes. By prising open the systemic closure in which Earth systems scientists – and a great many social scientists – are increasingly invested, geologists and geophysicists with a 'deeper' purview remind us of earthly life's fundamental exposure to an unbound, unassimilable exteriority. Such an opening, I want to suggest, draws the concept of resilience in directions other than Earth system science's ideal of well-managed planetary boundaries and elastic, adaptable human agents. But so too does it trouble the relegation of the resilient subject to the demands of neoliberal capitalism – or any other critical appraisal that would consign vulnerability and its enframings to a self-contained sphere of sociality.

Bodily vulnerability, planetary trauma

Acting as both disequilibriating shocks and sources of nourishing elements, periodic mass-eruptions of matter-energy from the inner Earth have increasingly been viewed by Earth and life scientists as stimuli of major evolutionary changes of direction. Currently dated at around 66 million years ago, the vast Deccan Traps flood basalt outpouring at the boundary between the Cretaceous-Paleogene periods is implicated in the demise of non-avian dinosaurs and the subsequent proliferation of birds, fish and mammals (Fortey 2005: 272–83). In turn, the monstrous Siberian Traps eruptions some 250 million years ago that mark that the Permian and Triassic boundary played a part in the die-off of an estimated 90% of the Earth's species – creating opportunities that eventually saw dinosaurs dominating terrestrial ecosystems. Still further back, Earth scientists have identified massive bursts of magma that spewed across the Columbia supercontinent around two billion years ago. Providing an abundance of bioessential elements as it gradually eroded, this mineral rich magmatic effusion has been linked to the rise of eukaryotes – micro-organisms with complex cell structures that are the ancestors of all multicellular life, including algae, plants, fungi and animals (Parnell et al. 2012; see also Clark et al. 2018).

If these are amongst the more momentous junctures in evolutionary history, so too is the permeability of the crust to the planet's seething interiority implicated in a multitude of lesser re-routings, including the volcano-strewn rifting of the eastern African continent that played its part in the emergence of the genus *Homo*. More than stimuli or excitations, the susceptibility of the surficial Earth to the forces of an unliveable outside is a reminder, to borrow from literary theorist Clare Colebrook (2010: 45), that '[n]o living body is the author of itself'. And in this way we begin to get a sense that, however vital a sustaining network of connections or entwinings might be in the enduring of stressful events, earthly life is conditioned by forces over which it has little or no influence.

Philosopher and literary theorist Gayatri Chakravorty Spivak, in a more general vein, describes this basic structural condition of openness to spatio-temporal otherness 'as the primordial wound of living-in-time' (Spivak 1999: 333). Or as cultural theorist Pheng Cheah would have it, likewise writing in a deconstructive register: '[i]t is precisely this internal vulnerability of any present being to alterity – its pregnancy with the movement of altering – that allows something to alter, change, or transform itself ... or to be changed, transformed, or altered by another' (Cheah 1999: 191). If such a logic begins to unsettle the opposition between vulnerability and resilience, to acknowledge a constitutive exposure or susceptibility along these lines is not to idealize a complete openness, for a being or system that lacks any capacity to regulate the exchange between itself and its outside is one that invites total dissolution (see Clark 2017: 10–11).

In *Beyond in the Pleasure Principle* (1920), his post-World War I essay on trauma, Sigmund Freud explored the tension between exposure to excessive stimuli and the establishment of a defensive barrier. He did not, however, restrict his analysis to psychic responses, and neither did he stop at the *human* body. In what he confesses to be 'far-fetched speculation' (Freud 1961[1920]: 18), Freud reflected on the way that all living organisms construct a skin, shell, or crust to modulate the potentially overwhelming forces of their environment:

> This little fragment of living substance is suspended in the middle of an external world charged with the most powerful energies; and it would be killed by the stimulation emanating from these if it were not provided with a protective shield against stimuli.
>
> *(Freud 1961[1920]: 21)*

In this process of self-defence, observes Freud, the individuating organism – like the similarly susceptible human psyche – must sacrifice some of its sensitivity by hardening itself at the zone of contact with its milieu. An experience can be considered 'traumatic' when stimuli or excitations from the outside break through this protective barrier (Freud 1961[1920]: 23). At the same time, Freud continues, the developmental process of erecting a boundary between self and world is itself experienced by the living being as a loss, a severance, a scarring – giving rise to deep-seated impulse to return to pre-individuated continuity with the outer world (Freud 1961[1920]: 30–1). This is much the same logic that deconstructive thinkers will later deploy: the idea that a 'decision' has consequences, that the cut – for all that it may be inaugural and generative – will bleed further on (see Spivak 1999: 332–5, fn 31, 33).

Remarkably, having proceeded from the battle-scarred or developmentally damaged human psyche to the generalized condition of the biological organism, Freud extrapolated still further. As if deferring making the final cut in his own schema, he gestured towards a constitutive tension between boundedness and exposure that extended all the way to the geocosmic scale (Freud 1961[1920]: 496–7). 'In the last resort', mused Freud, 'what has left its mark on the development of organisms must be the history of the earth we live in and of its relation to the sun' (Freud 1961[1920]: 32).

It has taken the better part of a century for the truly 'far-fetched' aspects of *Beyond the Pleasure Principle* to be fully appreciated. Embellished in conversation with Deleuze and Guattari's geophilosophy, Freud's abyssal extension of trauma was elaborated into quasi-fictional theory of 'geotraumatics' by philosopher Nick Land in the same millenium-closing year that saw the announcement of the Anthropocene. As Land puts it: `Fast forward seismology and you hear the earth scream. Geotrauma is an ongoing process whose tension is continually expressed – partially frozen – in biological organization ... Evolution presupposes specific geotraumatic outcomes' (Land 2011: 499). Developed by fellow philosopher Reza Negarestani in the following decade, the concept of geotrauma is systematized into a schema involving a succession of 'nested' levels of existence, each one at once vulnerable to and painfully severed from the forceful milieu that gives rise to it. 'Since there is no single or isolated psychic trauma (all traumas are nested)', observes Negarestani, 'there is no psychic trauma without an organic trauma and no organic trauma without a terrestrial trauma that in turn is deepened into open cosmic vistas' (Negarestani 2011: 1–2).

There is a great deal more going on here than metaphorical to-ing and fro-ing between the injuries afflicting human flesh and the rifting of the geobody of the Earth, though the profusion of figurations of organic-planetary wounding that we touched upon earlier are themselves suggestive of an intuition that goes deeper than wordplay. Both Land and Negarestani are talking about cuts in the real – consequential cisions in the stuff of existence whose legacy is an enhanced capacity for 'survival' – if that is not too organic a term – but whose price is the permanent scar of partition. While far from identical, psychic barriers to excessive excitation, the skin or husk of the organism that divides bodily interior from environment, and the planetary crust with its biospheric envelope that separates inner Earth from impinging cosmos, follow a similar structural logic. In this regard, Freud's mental apparatus 'flooded with large amounts of stimulus' (Freud 1961[1920]: 23) belongs to the same chain of nested traumatic irruptions as the flood basalts that periodically breach the Earth's crust. And in both cases – the maintenance of the 'membrane' – the negotiation between interior and exterior – is essential and ongoing; endlessly enabling and perpetually fraught (see Derrida 2005: Clark 2017a).

The Earth system science framing of resilience, as we have seen, is oriented towards pressures or excitations generated *within* the system. The sources of change to which Earth systems or ecosystems must respond arise out of the coupling of anthropogenic, atmospheric, hydrospheric and biospheric forces: a largely bounded set of interactions we might characterize as a kind of auto-affection. But this way of viewing terrestrial existence – according to the structural logic of traumatics – itself rests upon a decisive intervention. By severing the sphere or envelop of the outer Earth system from the perturbations of both the cosmos and the inner Earth, it becomes possible to imagine a certain self-authorship of life – which under Anthropocene conditions foregrounds auto-affective human life. In insisting that the generic category of the human or Anthropos is sub-divided, differentiated, rigorously re-socialised, 'progressive' social science advances rather than questions this logic of self-enclosure. Through their shared disavowal of a radical exteriority to the domain of liveability, I contend, Earth system science and critical social thought are ultimately complicit in the dream of collective, human-guided planetary authorship or 'governance'. However timely aspects of this vision may be, the cut upon which it depends – in the final instance – severs radical or revolutionary social intervention from the very 'excitations' of geo-cosmic existence that may be its most deadly weapon.

Resilence and the primordial wounds of living-in-geological-time

Attuned to the relentless dynamics of our planet's surface and near-surface, Earth system science promotes a concept of resilience befitting a world that pulses with potential for change – at every scale. There is much to agree with in the ensuing assertion that minimizing the shocks of the Anthropocene calls for every effort to avoid further degradation of vital systems along with the cultivation of new social support mechanisms and crisis-ready institutional forms. And neither should psychic resilience be too hastily dismissed as anti-social individualization. Freud, it is worth remembering, advanced his own version of resilience in the mental and somatic apparatus, though he was also quick to acknowledge its limits:

> In the case of quite a number of traumas, the difference between systems that are unprepared and systems that are well prepared ... may be a decisive factor in determining the outcome; though where the strength of a trauma exceeds a certain limit this factor will no doubt cease to carry weight.
>
> *(Freud 1961[1920]: 25–6)*

Freud is talking about *external* shocks or excitations here. With analogies to contemporary Earth system thinking, however, he also explored the possibility of psychic-somatic systems generating their own internal excitation – in which case he recognized that barriers erected by the organism to defend against threats from the outside would have implications for the way internally generated stimuli were dealt with (Freud 1961[1920]: 22–3). So while we should not underestimate the importance of the transformative possibilities that any complex system – psychic, somatic, ecological, geophysical – can generate though its own forcing and feedbacks, so too must we be attentive to the constitutive role of exteriority and to the lasting repercussions of mechanisms that have evolved to shield impingements from outside. It is in this regard that the speculative notion of geocosmic traumatics draws our attention to the succession of cuts or severances from which any actual system has composed itself – and to the ways in which it must work across or through the resultant 'scars' in order to maintain itself. In short, to paraphrase Spivak, there can be no resilience that is not a reckoning with the primordial wounds of living-in-geological-time.

Although reflection on suffering – the search for its meaning – might be a peculiarly human vocation, literary theorists Tim Matts and Aidan Tynan observe that '(s)uffering is, essentially, *not* a human problem, but primarily a *geological* one, a concern of the earth' (Matts and Tynan 2012: 102, 95). With their detailing of the upheavals that distinguish one stratum from the next and their meticulous archiving of life's perishability – it is the 'old school' hard rock geologists who beckon us into the wrenching depths of the Earth: the stratigraphers and paleontologists who remind us that our planet is at once a treasury and a cemetery of fossil-laden strata. Likewise, we can read the human body – any organismic body – as a repository of geological and climatic upheavals that have been ridden out. '[T]he time of the earth is recorded, accreted, knotted up inside us,' as philosopher Robin Mackay (2012: 20–1) puts it: a reminder that the traces of a fantastic voyage through the planet's abyssal depths are always already within us – or rather, that those traces are 'us'.

So too might the very thought processes by which our species strives to make sense of its predicament be seen as a series of attempts to cast lines across the rifts opened by an injurious Earth and cosmos. Is Morton's 'abyss whose reality becomes increasingly uncanny ... the more scientific instruments are able to probe it' the condition of the Anthropocene, we need to ask, or is it more generally the condition of human learning and thinking? For as Jacques Derrida would have it, reflecting upon the otherness that our very existence opens us up to: '[t]his incomprehensibility is not the beginning of irrationalism but the wound or inspiration which opens speech and then makes possible every logos or every rationalism' (Derrida 1978: 98).

For all its aura of novelty – and many of its techniques and concepts are indeed new – the Anthropocene thesis is not western thought's first sustained encounter with the perishability threatened by planetary upheaval. In the work of the most influential 18th and early 19th century European philosophers, what appears again and again – if we look past the veneer of enlightened self-assurance – is the anguish of dwelling on a violently discontinuous Earth. Fully engaged with the geological discoveries of their time, Kant, Hegel, Schelling and many of their contemporaries confronted the amassing evidence of life-extinguishing 'revolutions of the Earth' that were inscribed in the geologic strata (Clark 2017b: 217–19). Each thinker in their own way responded to the threat of the globe being 'dissolved into chaos', as Kant (1993 [1796–1804]: 66–7) put it, by seeking some enduring principle and schema that would insulate human freedom from a menacing geocosmic exteriority

In this regard, we might conceive of the physical threat of a revolutionary Earth as a primary incitement to the very idea of an autonomous, self-directed, social being: the discovery of deep, cataclysmic geological time as the wound or inspiration that propelled modern European thought towards the consolations of a bounded 'human' or 'social' science (see Clark and Yusoff 2017: 3–4). If this is the case, then every attempt of contemporary critical thought to contain and reclaim the shock of the Anthropocene in exclusively 'social' terms is a re-enactment of 18th–19th century geotrauma, a fortifying of the protective shield behind which the thinking of the social established its self-identity.

Is Earth system science, by this reasoning, a breakthrough in coming to terms with the inherent variability and volatility of our planet? Or is it the latest and most grandiose of western thought's successive efforts to bind and assimilate exteriority into a spherical totality: a final surge of the project of self-enclosure through which the cosmos is downsized and the domain of life amplified until they appear co-extensive and interdependent? These are questions we must also ask of any critical narrative that imputes such force to capitalism or EuroAtlantic modernity or colonialism that they end up mapping onto the planet without remainder.

While caution must be taken not to simply reduce Earth system science and its directives to a ruse of capital, we should also consider Negarestani's point that capitalism has proven a powerful vehicle for reaching out into geological (and incipiently, cosmological) depths – and drawing what was once extraneous into its own orbit. 'In binding the exorbitant register of exteriority', Negarestani observes, 'capitalism is able to present its dynamism as an intrinsic planetary system' (Negarestani 2011: 16). In this sense capitalism itself needs to be construed as a response to the greater geocosmic predicament, though it is a response that seeks always to appropriate and monopolize the geotraumatic impulse for its own interest – which ultimately means in ways that deepen and exacerbate life's exposure to the rifting of the Earth (see Negarestani 2011: 17)

Whether couched in terms of geotrauma – or any other approach that confronts the constitutive exposure of humans and other terrestrial life forms to an enabling and threatening exteriority – the provocation is one of how to find ways to meet a 'revolutionary' Earth and cosmos on its own terms. Any exploration of 'resilience', any means of extending our improbable and fantastic journey through the repeated upheavals of an exceptionally unstable planetary body, must pass through the scars of planetary, organismic and psychic upheaval and not over or around them. If this raises some fundamental doubts about critical tactics that aim to contain the incitements of Anthropocene geoscience in conventional social categories and concepts, too does it raise questions about the current wave of initiatives for 'governing' Earth systems that quickly settle on familiar strategies of transnational co-operation and institution-building (see for example, Biermann 2012; Wijkman and Rockstrom 2012: 174).

More than a problem of facing up to its inadequate attention to *socio-cultural* differences, the trouble with recent strategies to protect planetary boundaries is their ideal of smoothing the socio-political sphere into a seamlessly, inter-connected globality that insufficiently acknowledges its *geological* rifts, differentials and divisions (Clark 2016; 2017b). And in this way, the path towards resilience remains the extension and consummation of the existing *socius*, rather than the probing of the social field in search of the deep history of shock, damage, repression, deflection, adaptation and acclimatization out of which it has been assembled. For as Negarestani suggests, it is in the very experience or trace of exposure to the unassimilable forces of the Earth and cosmos that we are likely to find our most potent provocations, powers and tactics for effective change. 'The revolutionary subject restlessly searches for alternative syntheses or modes of traumatic inflection ...' he proposes. 'It improvises out of its traumas, or to be more exact, out of traumas which mediate between its regional horizon and the outside' (Negarestani 2011: 11).

Clearly, we are not going to be provided with a manual for step-by-step geotraumatic insurgency. We might, however, think in terms of a productive conversation with philosopher Elizabeth Grosz's call for to experimental 'involution' of terrestrial and cosmic forces: her invitation to find creative and expressive ways to calve off, isolate and downsize an otherwise overwhelming exteriority to a scale at which we can more safely work with these powers (Grosz 2008: 3; Grosz 2011: 38). Though in the process we ought to heed Grosz decidedly geotraumatic cautioning that 'art is also capable of that destruction and deformation that destroys territories and enables them to revert to the chaos from which they were temporarily wrenched' (Grosz 2008: 13). Here too we should consider geographer Stephanie Wakefield's (2017) injunction to inhabit the back loop of newly liberated matter-energy in the wake of ecological catastrophe, keeping in mind that the loop also spirals abysally through the succession of improvised reorganizations that have followed cosmic, geologic and evolutionary upheaval. And we need to think in terms of the great rifts gouged by Euro-modern colonialism in the physical and experiential worlds of others – though such deep and pervasive chasms of socio-eco-geotrauma demand their own full accounting (see Yusoff 2018).

With its awkward if productive tension between an ascendant Earth system science and an older stratigraphic geology, the scientific discourses of the Anthropocene serve at once to bind biogeophysical processes to the world-making efforts of our own species and to remind us of the inhuman abyss that yawns beneath every anthropic endeavour. How Anthropocene science and the related field of global change are constructing resilience reflects this equivocation, though the current trend seems to be favouring the assumption that the Earth system is contiguous or co-extensive with the spheres of human collective agency.

But Steffen's genre-defining assertion that 'the forcings and feedbacks within the Earth System are as important as the external drivers of change', I have been suggesting, can and should be read the other way round. As variations played on the theme of geotrauma remind us, and by sciences' own estimates, these external forcings have a head start on us of at least 13 billion tumultuous years. An uneasy amalgam of extrapolative readings of the natural sciences, pyschoanalysis, continental philosophy and genres of fantasy and science fiction, geotraumatics prompts us to think of resilience as a reckoning with scar tissue, an opening of old wounds that may help us to endure fresh injury, a sensitivity towards an inescapable exposure to new shocks and rumblings. It provokes imaginings of resilience that put 'coping' or 'adaptation' into confrontation with 'revolutionary' upheavals that belong as much to the Earth as to any recognizably social domain. What we might make of geotrauma as an incitement to terrestrial revolution remains to be worked through, as does the question of what further cuts made by geocosmic realities will make of us.

References

Bai, X., Van Der Leeuw, S., O'Brien, K., Berkhout, F., Biermann, F., Brondizio, E. S., ... Syvitski, J. (2016) Plausible and desirable futures in the Anthropocene: a new research agenda. *Global Environmental Change* 39: 351–362.

Benson, M. H. and Craig, R. K. (2014) The end of sustainability. *Society and Natural Resources*, 27: 777–782.

Biermann, F. (2012) *Earth System Governance: World Politics in the Anthropocene.* Cambridge, MA: MIT Press.

Brooke, J. (2014) *Climate Change and the Course of Global History: A Rough Journey.* New York: Cambridge University Press.

Cheah, P. (1999) Spectral nationality: the living-on [sur-vie] of the postcolonial nation in neocolonial globalization. In E. Grosz (ed.), *Becomings: Explorations in Time, Memory, and Futures.* Ithaca, NY and London: Cornell University Press.

Clark, N. (2016) Anthropocene incitements: toward a politics and ethics of ex-orbitant planetarity. In R. Van Munster and C. Sylvest (eds), *Assembling the Planet: The Politics of Globality Since 1945*. London: Routledge, 126–144.

Clark, N. (2017a) PyroGaia: planetary fire as force and signification. *Ctrl-Z: New Media Philosophy* (7).

Clark, N. (2017b) Politics of strata. *Theory, Culture & Society* 34(2–3): 211–231.

Clark, N., Gormally, A. and Tuffen, H. (2018) Speculative volcanology: violence, threat and chance in encounters with magma. *Environmental Humanities* 10(1): 273–294.

Clark, N. and Gunaratnam, Y. (2017) Earthing the Anthropos? From 'socializing the Anthropocene' to 'Geologizing the Social'. *European Journal of Social Theory* 20(1): 146–163.

Clark, N. and Yusoff, K. (2017) Geosocial formations and the Anthropocene. *Theory, Culture & Society* 34(2–3): 3–23.

Colebrook, C. (2010) *Deleuze and the Meaning of Life*. New York: Continuum.

Crutzen, P. (2004) Anti-Gaia. In W. Steffen *et al.* (eds), *Global Change and the Earth System: A Planet Under Pressure*. Berlin: Springer-Verlag, 2.

Davis, M. (1996) Cosmic dancers on history's stage? The permanent revolution in the earth sciences. *New Left Review* 217: 48–84.

Derrida, J. (1978) *Writing and Difference*. London: Routledge.

Derrida, J. (2005) *Rogues: Two Essays on Reason*. Stanford, CA: Stanford University Press.

Folke, C. (2004) Enchancing resilience for adapting to global change. In W. Steffen *et al.* (eds), *Global Change and the Earth System: A Planet Under Pressure*. Berlin: Springer-Verlag, 287.

Fortey, R. (2005) *The Earth: An Intimate History*. London: Harper Perennial.

Foster, J. B., Clark, B. and York, R. (2010) *The Ecological Rift Capitalism's War on the Earth*. New York: Monthly Review Press.

Freud, S. (1961[1920]) *Beyond the Pleasure Principle*. New York and London: W.W. Norton.

Grosz, E. (2008) *Chaos, Territory, Art: Deleuze and the Framing of the Earth*. Durham, NC: Duke University Press.

Grosz, E. (2011) *Becoming Undone: Darwinian Reflections on Life, Politics, and Art*. Durham, NC: Duke University Press.

Hamilton, C. and Grinevald, J. (2015) Was the Anthropocene anticipated? *The Anthropocene Review* 2(1): 59–72.

Kant, I. (1993[1796–1804]) *Opus Postumum*. Cambridge: Cambridge University Press.

Land, N. (2011) *Fanged Noumena: Collected Writings 1987–2007*. Falmouth: Urbanomic.

Latour, B. (2014) Some advantages of the notion of 'critical zone' for geopolitics. *Procedia* 10: 3–6.

Lenton, T. (2016) *Earth System Science: A Very Short Introduction*Oxford: Oxford University Press.

Lövbrand, E., Beck, S., Chilvers, J.*et al.* (2015) Who speaks for the future of Earth? How critical social science can extend the conversation on the Anthropocene. *Global Environmental Change* 32: 211218.

Mackay, R. (2012) A brief history of geotrauma. In E. Keller, N. Masciandaro and E. Thacker (eds), *Leper Creativity: Cyclonopedia Symposium*. New York: Punctum Books, 1–36.

Masco, J. (2010) Bad weather: on planetary crisis. *Social Studies of Science* 40: 7–40.

Matts, T. and Tynan, A. (2012) Geotrauma and the eco-clinic: nature, violence, and ideology. *Symplokē* 20(1–2): 91–110.

Morton, T. (2012) Ecology without the present. *The Oxford Literary Review*, 34(2): 229–238.

Negarestani, R. (2011) *On the Revolutionary Earth: A Dialectic in Territopic Materialism*. Paper presented at Dark Materialism, Kingston University, Nat Hist Museum London, 12 January 2011. Online at: http://s3.amazonaws.com/arena-attachments/77505/20457-on_the_revolutionary_earth.pdf?1360838364 (accessed 1 November 2018).

Parnell, J., Hole, M., Boyce, A. J., Spinks, S. and Bowden, S. (2012) Heavy metal, sex, and granites: crustal differentiation and bioavailability in the Mid-Proterozoic. *Geology*, 40: 751–754.

Self, S. and Rampino, M. (2012) The crust and the lithosphere. *The Geological Society*. http s://www.geolsoc.org.uk/flood_basalts_2 (accessed 1 November 2018).

Spivak, G. C. (1999) *A Critique of Postcolonial Reason: Toward a History of the Vanishing Present.* Cambridge, MA: Harvard University Press.

Steffen, W. and Lambin, E. (2006) Earth system functioning in the Anthropocene: human impacts on the global environment. *Interactions between Global Change and Human Health. Pontifical Academy of Sciences, Vatican City, Scripta Varia* 106: 112–144.

Steffen, W., Leinfelder, R., Zalasiewicz, J., *et al.* (2016) Stratigraphic and Earth System approaches to defining the Anthropocene. *Earth's Future* 4(8): 324–345.

Steffen, W., Sanderson, A., Tyson, P., *et al.* (eds) (2004) *Global Change and the Earth System: A Planet Under Pressure.* Berlin: Springer-Verlag.

Than, K. (2007) Mission to study Earth's gaping 'open wound'. *Live Science.* http://www. livescience.com/1317-mission-study-earth-gaping-open-wound.html (accessed 20 October 2018).

Yusoff, K. (2018) *A Billion Black Anthropocenes or None.* Minneapolis, MN: Minnesota University Press.

Wakefield, S. (2017) Inhabiting the Anthropocene back loop. *Resilience: International Policies, Practices and Discourses* 6(2): 77–94.

Wijkman, A. and Rockström, J. (2012) *Bankrupting Nature: Denying our Planetary Boundaries.* Abingdon: Routledge.

Winchester, S. (2004) *Krakatoa: The Day the World Exploded.* London: Penguin.

Zalasiewicz, J. (2008) *The Earth After Us.* Oxford: Oxford University Press.

Zalasiewicz, J., Steffen, W., Leinfelder, R., Williams, M. and Waters, C. (2017) Petrifying earth process: the stratigraphic imprint of key earth system parameters in the Anthropocene. *Theory, Culture & Society* 34(2–3): 83–104.

10

MORE OF THE SAME?

Life beyond the liberal one world world

Stephanie Wakefield

Introduction: twilight/dawn

While in the 1990s the 'end of history' (Fukuyama 1992) or the global triumph of liberalism's 'one world world' (Law 2015) may have seemed plausible, recent years have led to an almost universal certainty that things are coming apart. Hurricanes have sent record storm surge heaving over coastal walls, and cities once thought unshakeable are now portrayed as fragile places imperiled by myriad risks – rising seas, heat waves, technical failures – that threaten infrastructural systems. Decades of neoliberal revanchism and cybernetic governance have dismantled both the modern subject and the grounds upon which it was built. Social conflict, rather than being pacified, has proliferated and mutated into forms barely recognizable via past political rubrics, including the rise of tyrannical social media-based battles amidst an increasingly fractured, if not tribalistic landscape. In the words of its own proponents, the conditions of liberal civilization seem to be coming undone, a sense buttressed by the growing number of scientists prophesizing the end of civilization due to climate change (Ogilvy 2017; Sullivan 2019; Torres 2017). Thus, from the present vantage point, far from tightening, liberalism's hold on the world looks instead to be severely weakening.

While emerging from the geological and earth system sciences (Crutzen 2002) and popularly understood as a purely environmental matter, the Anthropocene offers an important label for this situation. The Anthropocene – the Epoch of the Human – has received extensive criticism for its invocation of a single figure of Man (or The Human or Anthropos), which authors have taken to task variously for what is seen as its erasure of race and gender difference (Grunsin 2017; Haraway 2016; Yusoff 2019) or its elision of the fact that the destruction now wrought by 'humanity' is in fact caused by the actions of a very small percentage of wealthy humans (Malm and Hornborg 2014; Moore 2015). While such arguments are important and contain important truths, in my view its invocation of a single definition of human life is the Anthropocene thesis's greatest virtue (Wakefield 2014a; 2014b). For this way of thinking about and molding life within one frame is that of modern liberal regimes, which the Anthropocene thus refers to as an historically specific – dated and finite – strategy for approaching human being. More specifically, the Anthropocene names the liberal project of defining and enforcing life as a 'one world world' in order to call all of this a *failure*, evidenced in the degradation of natural environments and human subjectivity alike. Thus the Anthropocene – and what I have elsewhere called our current place in its 'back loop' (Wakefield 2018a) – provides a name for the liberal way of life as one finds it today: a sinking ship increasingly taking on water from all sides.

The term's critical usefulness does not end there. Instead of heralding the building of a one world world – processes congruent with the Anthropocene's 'ascendant' phase – the moment of naming the Anthropocene is one of confusion, chaos, and the potential for transformation. Entering into the Anthropocene need not only be understood in terms of listicles of destruction wrought by a homogeneous 'human geological agency' but ought to be understood as a name for the twilight of liberalism, its single world order and fictive kinships coming undone and the opening to other possibilities that this unraveling permits.

As old grounds give way, in government, design, and planning as much as in critical theory and art, statements proliferate to the effect that the Anthropocene 'changes everything,' marking a rupture point when now-outdated rubrics must be thrown out, imagination and new ideas are needed, and experimentation is seen as the modality of the day (Klein 2007; Tsing 2016). But these declarations require more critical investigation. What practices and forms of life are being forwarded in such efforts? Moreover, despite the ubiquity of such efforts, are we actually seeing anything new or simply a new façade of liberalism? Taking the end of liberalism seriously and the possibilities now open to us is perhaps the key wager of our time. I explore this ground in the following pages first in two distinct fields – resilience practice and critical theory's rising notion of a 'pluriverse' – both of which forward calls for imagination and the new, and indeed take such calls seriously, yet I argue ultimately deliver more of the same. To conclude I explore other possibilities for life beyond the false choice between maintaining liberalism's one world world or surviving its end.

Governing liberalism's end

Faced with their own conditions of existence in disarray and the question of how to maintain their social order – summed up in New York City's former mayor Michael Bloomberg's (2012) statement, 'we have to do a better job not only keeping our networks up, but keeping our markets and businesses open, come hell or high water' – cities have begun to experiment with new modes of governance. In recent decades resilience has risen to the forefront as the answer to this challenge, quickly becoming the watchword and strategy of such liberal regimes in peril (Brown 2014). Defined variously as the ability to 'stay afloat,' 'bounce back,' or even 'bounce forward' amidst turbulence, resilience has become the mantra for the Anthropocene and its 'new normal' of rising seas, natural disasters, and infrastructural fragility.

Resilience is seen as offering the promise of the radically new, able to solve past failures. For some, resilience promises to heal the nature/culture divide by uncovering nature's potential or reconnecting urban dwellers with nature, each other, or earth systems (Lourie Harrison, 2013) – as if imagining a fully interconnected city could somehow soothe the latter's volatility. In ecology and systems thinking, fields from which the concept itself emerged and where it is today much discussed, resilience is promoted as a corrective to outdated 'stability' models with their notion of ecosystems as single steady state and equilibrium and efforts to manage them by preventing disruption or change (Folke 2006). For resilience thinkers such as Carl Folke (2006), this is now an 'old' perspective that 'provide[s] little insight' (2006: 253; 256) and is in fact an 'alienated' 'pathology' actively preventing progress and increasing vulnerability. In its place, resilience thinkers posit what they see as a more realistic, accurate, and contemporary view of the world as made up of dynamic, fundamentally unstable, coupled, nonequilibrium social and ecological systems (Berkes and Folke 1998). One must now, as Folke puts it, 'learn to manage by change rather than simply react to it … uncertainty and surprise is part of the game and you need to be prepared for it and learn to live with it' (Folke 2006: 255). Though its definition and approaches to its management are seen as works-in-progress, amongst ecologists resilience is seen as having 'universal applicability' and is an 'inherent property of systems' that today must be developed (Fath et al. 2015: 1).

Across diverse discourses, one thing is emphasized: resilience government must take a different tack than techniques past. It cannot come from top down onto landscapes conceived as blank slates, but requires collaboration between diverse actors and communities and must incorporate local environments. In contrast to security, discipline, or outright control, resilience has been heralded as offering particular promise for a much broader range of voices due to its recognition of Anthropocene dislocations and calls for a thorough shift in thought and design away from dualist thinking and toward nonequilibrium entangled eco-cybernetic systems approaches (Folke 2006; Gunderson and Holling 2002). Cast in terms of innovation, 'bouncing forward,' risk-taking, and as a paradigm breaking with obsolete ways and willing to throw out old models, resilience is portrayed as cutting edge and cool – rather than as a project of urban security and governance – said above all to require experimentation. The latter perspective is endemic to resilience thinking. As resilience founder C. S. Holling himself puts it, when faced with a situation of such 'unknowns' as constitute the Anthropocene, instead of rehashing old models one should not 'try to plan the details, but invent, experiment' (Holling 2004: 7; Gunderson and Holling 2002).

Such experiments are nowhere more common than in the realm of urban climate change governance. In the wake of recent hurricanes – and faced more broadly with the matter of rising seas and floods undermining the geological and infrastructural ground on which liberal regimes are built – cities are rebranding as laboratories for resilience, becoming first responders for experimentation in new techniques to manage liberal life. In places like New York, dominated through the 20th century by large-scale engineering and Robert Moses-style top-down architecture (Gandy 2003), landscape ecology and systems thinking now predominates. Instead of carving massive highways across neighborhoods or plopping down Army Corps of Engineer cement sea walls on landscapes seen as empty canvases, the resilient approach means experimenting with the use of aquatic life such as wetlands and oyster beds to manage storm surge, and tapping community groups to generate plans alongside celebrity architects in design charrettes. High profile competitions are launched to attract world-class design talent – think REM Koolhaus, Bjarke Ingels Group, an array of Harvard University Graduate School of Design alumni – and to recruit 'those willing to entertain and support unconventional solutions' (Jacobs 2017). In cities where the resilience imperative looms large, events, symposia, and working groups dedicated to reimagining aqua–urban futures are commonplace events, with city governments like Miami Beach embedding artist-in-residences to assist in framing their resilience. Alongside such projects, a redux of 20th century engineering mania can also be found amongst scientists and engineers exploring options to elevate streets and whole cities (Wakefield 2019).

Despite its emphasis on novelty and innovation, within the majority of resilience experiments what is posed is not a matter of transforming how we live, but instead, how to *govern* the dislocations of the Anthropocene? More specifically, how to maintain liberal regimes amidst and despite challenges to its hegemony, stability, and eternality, amidst the very rising seas, droughts, and floods liberal regimes themselves engender (Wakefield and Braun 2018)? How to keep networks, markets, and businesses open, come hell or high water? As Gareth Edwards and Harriet Bulkeley (2018) have argued of urban climate experimentation more broadly, resilience experimentation represents a new mode of government, not an exception to it. Now composing a biopolitical *dispositif* – the purview of which is everything from insurrections and social relationships to infrastructural failure and natural disasters – resilience has been added to a suite of strategies which include discipline and security (Wakefield and Braun 2014; Braun 2014). Resilience's power brokers have not been shy about making such conservative priorities public. As former-Rockefeller Foundation president and resilience advocate Judith Rodin (McDonnell, 2012) succinctly summarized it: 'This isn't just climate-related. We're not just thinking about hurricanes or floods; we're really thinking about any vulnerability to the system that could take it down, and how to build against that.'

Like past modes of liberal governance, resilience's key goal according to both its thinkers and political proponents is to attenuate and govern disruption in order to maintain the identity of existing systems. However resilience does so in new ways. As a development and transformation within a longer history of liberal governance amidst a time of the latter's growing crises, resilience responds to a 'crisis' or 'mutation' erupting in government's own problematic (Foucault 1984) – that of modernist control-based governance increasingly seen as outdated, the cause of climate change itself (Evans 2016; Dalby 2013). Far from denying the dislocations of the Anthropocene, resilience thinkers acknowledge them directly and call for overhaul in response to what they are said to reveal (human/nature entanglement, for example). In responding to the problematization of past governance approaches, resilience maintains systems identity and function, *through* change, not by preventing it. New governance structures or design paradigms are welcome – indeed necessary, according to the problematization of modern governance as it has been posed – in so far as they work toward maintaining the one world world's critical system parameters.

Resilience is thus fundamentally characterized by change: change in approaches to design and infrastructure, but also environmental changes, like global warming, which it sees as inevitable, and allows, rather than prevents. But resilience designs in no way change the way we live. They may *modulate* liberal subjectivity – transforming subjecthood into a form of crisis management and survival, vulnerability and worry (Evans and Reid 2014). At root however their goal is to maintain existing systems – the one world world of liberalism, with its infrastructural supports and so on – via transfiguration and transvaluation (Nietzsche 2010) of modern governmental techniques and truths. From this perspective experimental projects are means to 'navigate' or safely 'pass through' the turbulence of the Anthropocene, in order to endure and stay afloat during it (Fath et al. 2015). The goal of resilience, then, is to make liberal regimes into *survivors*, in the sense of surviving turbulence and change, via ever-changing design and infrastructural approaches, while maintaining regimes' essential identities within buffer zones. In the same breath radical transformation is promised – i.e. resilience experiments that overturn modern ontologies – while what is simultaneously guaranteed is the worsening of conditions and a future of inevitable crisis.

Resilience infrastructure must therefore be understood as the substrate of a liberal regime promising neither redemption nor progress but only survival of its own existing conditions amidst their ongoing breakdown. A mode of governing liberalism's own end, one devised amidst the *realization* that this end is occurring (a realization and fact which the Anthropocene names), resilience has the effect of tethering innovation and imagination to the maintenance of existing economic, social, and political relations. Beyond sea walls and wetlands most powerful of all resilience's techniques is its ability to conflate continuation of human life with the continuation of the specifically liberal way of life – to portray the two as necessarily synonymous – and moreover to portray the maintenance of both as the object and goal of each and all. That is, to make populations identify their own needs and desires in its own continuation (such can be seen in the taking up of resilience as motto for community groups and movements).

Rather than changing life, as a mode of government, resilience functions *to ward other ways of living off*. Adapting to changing conditions so as to keep all other things the same, resilience allows events to occur, in order to absorb, lessen, or incorporate them, so that existing social, economic, and political systems continue (Wakefield and Braun 2018). Change seems to be underway. Old models are rejected, exchanged for new more dynamic ones. Nature, no longer portrayed as a dump or object, is incorporated as a partner and infrastructural being, with urban design 'reconceived,' as Ross Exo Adams puts it, 'as a kind of benevolent inter-locutor in an otherwise "natural" process' (Adams 2014: 26). Imagined in AUTOCAD renderings, the same 'natural' vision of urban life forwarded is pro-tected by Wall Street-hugging sea walls, operational alongside fiber optic cables, pipelines, and logistical networks of Amazon and Google, BP, and Shell. Behind these walls and suspended amidst infrastructural networks, populations are buried in fear, debt, and dreams of the end. The possibility of imagining or creating other worlds disappears. Crisis-managing infrastructure becomes the dominant rubric through which all spheres of life are seen. Societies on lockdown, art, culture, and basic details of life become preparation for impending disasters seen as omnipresent and impossible to prevent, in defense of a way of living which has already been deemed a failure. As described by Adams, resilient urbanism thus constitutes a 'sinister' paradigm in which what is actually '"sustained" is, at best, a process without beginning or end, and whose only promise is the managed expansion of the urban regime' (Adams 2014: 27) – a single, global urban expanse without exterior amidst the naturalization of environmental collapse as a shared, inevitable global condition.

From liberalism to the pluriverse

Where resilience sees Anthropocene upheavals as necessitating new experiments in governmental techniques in order to maintain the liberal one world world, for other perspectives, particularly those found in critical theory, the Anthropocene marks the *end* of this unidimensional parameter for life, and the *opening* of other possibilities. Recently connected to this perspective is a growing literature calling for attention to the emergence and construction of a 'pluriverse.' Pluriverse thinking stems from efforts in the 1990s by diverse scholars based primarily in decolonial Latin American studies, bridging science and technology studies, indi-genous studies, and political ecology, around the project of 'political ontology' and a 'refusal to be captured by modernity' (Blaser 2013: 550; Mignolo 2002; 2007). Pluriverse theory is based on the key premise that modernity constituted what John Law (2015) has called a 'one world world,' a world which sees itself as unitary and universal and posits itself as the only world to which all other worlds must abide or be 'sentenced to disappearance in the name of the common goods of progress, civilization, development, and liberal inclusion' (de la Cadena and Blaser 2018: 3). Authors associated with this thinking have traced how the universalist project of liberalism – to order reality into a set of visible, manageable categories, based on particular ideas of a human-nature separation, humanism, profit, subjecthood, and exploitation – was imposed on the world in the colonial and post-colonial contexts (Escobar 2008). Consider here how liberal regimes sought the expansion of this single image of life via diverse strategies and techniques to secure societies inclusive of warfare, peacetime politics, everyday disciplines, infrastructures, extraction, and primitive accumulation, as much as mental frameworks and the imposition of ideologies of the common good (Foucault 1979). All require the concomitant elimination of ways of living deemed undesirable or 'illiberal' in both war and nonwar time (Reid 2005), replacing the many *nomoi* of the earth with a single *nomos* (Mignolo 2015).

In the face of the one world of liberal modernity, thinkers in this field argue for the existence of many worlds – a pluriverse, rather than a universe (Escobar 2008). Instead of reducing diverse forms of life to relativist categories like 'culture' – where differences between cultures are relative from the liberal omniscient, non-relative viewpoint (Blaser 2013: 550) – and extending anthropological work on the ontologies seen as inherent in indigenous practices (e.g., Viveiros de Castro 1998; Descola 2005), thinkers associated with the pluriverse project see liberal modernity as only one way of 'worlding' (Blaser 2013) or making worlds – entirely different realities – amongst a host of others. Rather than a single reality or world on which different cultures maintain diverse views, worlds are understood as enacted in practice, via the techniques with which people engage their environments and each other, and the diverse ways they understand these relations and the nature of reality itself (Burman 2017). Some pluriverse thinkers' point of departure is that there have always been other worlds, which have been subjugated or buried by the bulldozer of liberal universality. For others pluriversal thinking is conceived as a prospective project to be enacted, a 'delinking from the Western universal' (Mignolo 2018) or a 'de-noming,' overcoming the regulation of its institutions, laws, and knowledges (Mignolo 2015).

The pluriverse concept however has taken on particular purchase in association with the Anthropocene, which authors see as naming the crisis of hegemony of liberal modernity's ontological assumptions, where worlds and ontological conflict now have 'unprecedented visibility and potentiality' (Blaser 2013: 548; Mignolo 2018). Recent works from longtime pluriverse thinkers – *A World of Many Worlds,* a collection edited by Marisol de la Cadena and Mario Blaser, and Arturo Escobar's *Designs for the Pluriverse* – situate this proliferation of worldly possibility as specifically occasioned by the Anthropocene. Here the premise is that not only is the one world world a fiction – because other ways of living have always and continue to exist – but that today this fiction is increasingly coming undone. According to Escobar, the crisis of the Anthropocene is 'the crisis of a particular world or set of world-making practices, the world that we usually refer to as the dominant form of Euro-modernity (capitalist, rationalist, liberal, secular, patriarchal, white, or what have you)' (Escobar 2018: xx). Write de la Cadena and Blaser, 'the moment of the realization of the destruction of the Earth, the current historical moment can be one when people reconsider the requirement that worlds be destroyed' (de la Cadena and Blaser 2018: 4). In this perspective, instead of representative of Life, liberal modernity is only one way of life, one coming to a close. Emerging from the end of the 'one world world' is now said to be a 'world of many worlds,'[1] inclusive of the many forms of life and worlding practices that were slated for destruction with the West's conquest as well as those constituted in 'transitional' experiments underway now amidst Anthropocene pressures (Escobar 2018).

What does the pluriverse emerging in the Anthropocene look like? Drawing on a series of paradigmatic ethnographic examples including 'human practices with earth beings and with animal spirits that populate worlds,' and 'a mountain that is also an earth being and forest animals that are also spirit masters of their worlds' (de la Cadena and Blaser 2018: 12), de la Cadena and Blaser, for example, call for attention to 'presences that are or can be but do not meet the requirements of modern knowledge and therefore cannot be proven in its terms' (2018: 14). While from Western modernity's perspective mountains and forests appear as resources, for other practices, in this case the indigenous, they appear as 'persons' (2018: 5). In a similar vein, Escobar (2016) traces the emergence of other worlds within 'onto-logical struggles' to defend mountains, river, territories, or cultures against encroachment of the one world world, for example against gold mining industries in La Toma, Cauca, (2016: 15), as well as emerging 'transition' movements based in the Global North around 'post-growth, post-materialist, post-economic, post-capitalist, and post-dualist' and in the South as 'post-development, post/non-lib-eral, post/non-capitalist, and post-extractivist' (2016: 25), noting the inability of mainstream and Left commentators, steeped as they are in liberal modern thinking, to perceive what is actually emerging in such struggles. Other exemplary cases of pluriversality cited in these works include a collaborative effort joining modern mathematics and Yolngu Aboriginal methods of measurement in the Garma Maths Curriculum in Australia (Strathern 2018); the European Anti-GMO movement (Stengers 2018); the diverse worlds of 'the people of Pachamama' (Viveiros de Castro and Danowski 2018); and the 'aquatic space' of the Colombian Pacific (Oslender 2018).

Despite their emphasis on heterogeneity, in describing these examples and through them the pluriverse itself, authors often portray their pluriverses in sur-prisingly uniform ways. The pluriverse, it is said often, is made up of worlds said to eschew modern human/nonhuman binaries in favor of nondualist, entangled life upholding the importance of the nonhuman, with the possibility of encounters other than of a subject and an object promoted (de la Cadena and Blaser 2018: 10; 4). Particularly emphasized in Escobar's (2016; 2018) work is the matter of non-individualism and collectivism: the pluriverse in his account is made of 'collectives' 'which do not make themselves through [a nature/culture] divide' (2016: 15–16) and are 'communal,' having 'nothing to do with individual autonomy' (2016: 5). Such 'relational ontologies' as Escobar calls them are made of entanglements with a 'rhi-zome logic' to them – with tides, the moon, 'minerals, mollusks, nutrients, algae, microorganisms, birds, plant, and insect' (2016: 17) – in which 'things and beings are their relations, they do not exist prior to them' (2016: 18). Becoming is the norm, not the exception. 'These worlds do not require the divide between nature and culture in order to exist – in fact, they exist as such only because they are enacted by practices that do not rely on such divide' (Escobar 2016: 18). The argument in these works is that such modes of life are necessary to combat and move away from the one world world and its politics of individualism, private property, and nature/human separa-tion – a metaphysics seen as the ultimate cause of Anthropocene destruction. And yet, in forwarding a conglomeration of worlds each said to be entangled, nondualist, noncapitalist, non-individual, and privileging nonhuman actors and collectivism, pluriverse thinking inadvertently produces an equally homogeneous picture of life and its possibilities.

What to make of this homogeneity? Rather than opening up possibilities for life, could it be that by limiting the pluriverse to forms of life deemed desirable by analysts, despite its own intentions this discourse is instead closing possibilities down?[2] One way to approach this question is to note at a basic level that connectivity, relationality, and entanglement are equally the definitions of life forwarded by cybernetics, resilience, and systems thinking – and the models in which life has been materially reshaped over recent decades (thus dismantling the liberal subject in practice for some time). Might it be that instead of challenging the central beliefs of liberalism these discourses in fact provide new ways of forwarding what has become its central ideological and technological core? This argument resonates with that made by David Chandler and Julian Reid (2018), that such accounts are consistent with the images of life forwarded by resilience and contemporary neoliberalism, images increasingly discovered by resilience proponents in indigenous lifeways, which, along with nonhuman natures, are held up as superior ways of being capable of overcoming Western errors (thinking subjecthood as possession, individualistic superiority over and outside nature, subject/object binaries). Government promotes learning from indigenous lifeways, they argue, not for the benefit of indigenous peoples but as part of a search for new modes of managing populations, *within* the constraints of neoliberalism, not beyond them, and rather than offering emancipation constitutes 'a new form of neoliberal governmentality, cynically manipulating critical, postcolonial and ecological sensibilities for its own ends' (Chandler and Reid 2018: 252).

In a similar vein, these images resonate more broadly with a growing consensus amongst critical thinkers around what life in the Anthropocene should look like (Wakefield forthcoming). Here again, however, while emphasizing imagination, across these narratives one finds remarkably *similar* portrayals of the possibilities for life. In place of 'outdated,' 'parochial' (Tsing 2016: vii) modern ways of thinking and being, a new vision of life is forwarded, proclaiming universal definitions of the meaning of human being based in anti-humanism and messy entanglement in earths systems (Latour 2017; Grusin 2017; Haraway 2016; Hamilton 2017; Instone 2014; Tsing et al. 2017). While posed as a critically minded new political paradigm for the Anthropocene, such critical narratives often simply make explicit what is only implicit in resilience, forwarding images of life as insecure, entangled with, and hostage to volatile earth systems. Thus do governmental discourses and critical discourses converge in positing both similar lessons and similar definitions of life in the Anthropocene (Wakefield 2018b).

But perhaps more vexing than its proximity with governmental discourses is the more basic way in which the pluriverse, like resilience, is forwarded as a specific, *single* image of life (one moreover which is normative). Despite claims that something totally new is needed, what we find in both cases is more of the same; in resilience's case, extreme creativity and experimentation all toward managing existing social conditions of work, debt, profit, backed up by an unending state of emergency, ever-multiplying and expanding network of security techniques, social control, and police measures. In critical theory's case, more writing that assumes the meaning of thinking is to instruct readers in how to live, to provide universals to define life, alongside elite institutions and politicians telling 'us' that 'nothing else is possible,' that 'there is no alternative,' other than the continuation of the same. But can thinkers both announce the plurality of worlds and lay down parameters for their being?

It has been the method of liberal regimes since their inception to decide which ways of living are appropriate and which are not and, regarding the former, to mold individuals and populations in this image as much as, when faced with the latter, to eliminate them by whatever means necessary. Indeed the imposition of a single definition of life – and concurrent denial of the possibility that being is not a fact but a *question* – has long been the purview of liberal regimes but it is not limited to them. From Christian morality to Western metaphysics writ large, the naming of proper being or a single true world located somewhere above, beyond, or below life itself has equally been seen as the task of self-imagined gatekeepers of the real for some time. Yet just as the search for Christian morality over time appeared as an untruth, so now do diverse commentators see liberalism's will to truth as a fable. Such is nihilism as Nietzsche (1967: 7) defined it: when the long line of transcendent values floating above or beyond the world are devalued, yet one still cannot believe in the world or one's *own* values.

As Martin Heidegger (2002) saw it, despite aims of going beyond nihilism and metaphysics, Nietzsche's efforts to do so via a thinking of being as becoming remained trapped within metaphysics and nihilism itself. This was because, in Heidegger's view, rather than getting rid of the determination of being, Nietzsche still defined and valued it albeit by *reversing* it: what was God and the supersensory became will to power and the sensory, Being became becoming, etc. In this way, Heidegger argued, Nietzsche's attempt at transvaluation represented not an overcoming, but the completion or perfection, of metaphysics and nihilism. The ultimate critique which Heidegger made of Nietzsche was that, by still seeking to determine the meaning of being, Nietzsche continued to avoid the *question* of being, which was Heidegger's way of saying that there simply is no definition of what being is. We do not know in advance, outside the concrete practices through which people create their worlds.

Whether or not one agrees with Heidegger's assessment of Nietzsche's project – there are excellent reasons to believe the former willfully misunderstands the latter – his argument is a useful light in which to read resilience qua contemporary mode of liberal government as well as the perplexing problem of the homogeneous pluriverse. Resilience recognizes the exhaustion of liberal humanism qua Truth – an exhaustion announced by the Anthropocene, the epoch of liberal Man declared at the moment of Man's dethroning as catastrophe and failure – and responds to this with a new definition of liberal life: insecure and complex adaptive systems and human-nature entanglement. Pluriverse thinkers, on the other hand, despite their emphasis on the ontological plurality occasioned by the *breakdown* of liberal defi-nition of life – and despite their important recognition of the fact that waging war against the question of being is not only a philosophical matter for dusty library books but rather a political one – portray their pluriverses in surprisingly uniform ways as well, forwarding a conglomeration of worlds each said to be entangled, nondualist, noncapitalist, non-individual, and privileging nonhuman actors and collectivism. In both cases the *opposite* or *reversal* of what are seen as modernity's erroneous ways (separation of humans from nature, distancing from the natural world) are posited: entanglement with nonhuman beings not only as a possible way of understanding ones relationship to the world, but moreover as *the* relationship to the world; 'commons' and 'commoning' in opposition to property and individu-alism; etc. From resilience to pluriversality, in both cases the result is that a similar story is actually told: there is only one world. While resilience explicitly seeks to uphold a single world, unfortunately for pluriverse thinkers a single image of life is again forwarded, albeit by thinkers seemingly committed to the opposite of such unitary thinking. In this way the *question* that is being – the singular irreducible quality of life – is still forgotten. In thinking they know what being is and that they can determine it for others, each seemingly opposed discourse thus seeks the annihilation of being as a question.

War and other textures of relation

If this actually is the end of liberalism's one world world, are there not far more possibilities present to us now, for what life may be? Possibilities beyond main-taining liberalism beyond all sense, as does resilience? Possibilities beyond those which are defined merely in opposition to what are seen as the old world's cate-gories and metaphysics? What's more, possibilities beyond those defined in advance for others by academic thinkers? After all, if we take seriously the idea that liber-alism's single world order is unraveling, a thesis at the heart of pluriverse discourse, are there not more possible worlds? What of other ways of life emerging now?

Here think for better or worse of searches being carried out by diverse people for new practices and ways of explaining their life experience, from new and old fundamentalisms, prepping, fitness, and cultural movements (Wakefield forth-coming; Kelly 2016; Bennett 2009), to 'new tribalisms' both IRL and 'memetic' digital tribes (Wakefield 2017; Sullivan 2019; Junger 2016; Ronfelt 2007; Limberg and Barnes 2018; Burrows 2018; Nagle 2017). With the loss of credibility of gov-ernmental regimes as well as universal sciences, with media increasingly seen as promoting biased stories and not the objective facts of an Objective World as it Exists, here we may also file the rise of 'post-truth' and its critique of expertise and media objectivity (Marres 2018; Neimark et al. 2019). Such have democratic and anti-democratic, left and right, 'pro' and 'anti' capitalist expressions. Others do not fit in any of these boxes. What, one might add darkly, of the array of lifeways emerging in America's 'hinterland' (Neel 2018), such as those associated with the conservative Patriot Movement or Oath Keepers, which seek creation of local power bases and autonomy from the federal government in response to social and economic crisis? Each of these examples speaks to worlds which exist and are practiced today in direct response to the social, economic, and political dislocations of the Anthropocene, but they do not conform to the images of entangled, anti-human life said normatively to constitute it. What then is their status? Do they not exist in the pluriverse? Not qualify as a world?

Pluriverse thinker Andres Burman (2017; 2016) argues that hegemonic liberal epistemologies disallow and indeed suppress any other knowledges emerging from other lifeworlds and practices, such as those of the Aymara in the Bolivian Andes, leading such lifeworlds to become what Burman, with a twist on Fanon (1961), calls 'damnés realities': a 'subalternized' 'Other' relegated to a 'zone of nonbeing' (Burman 2017: 932). Might it be that the aforementioned range of worlds emergent today are subject to the same fate, albeit via relational-entanglement-collectivity discourse, becoming damnés realities of their own, 'lifeworlds that are not allowed to be, the lifeworlds that are censored by the coloniality of reality' (Burman 2017: 932)?

If one takes the idea that old world is ending seriously, this opens up much broader horizons. The end of the one world world and its fictive kinships is not the time for reasserting ever new definitions for what life should be, but for reaching out into the infinite range of what we and others might make it. Just as scholars such as Clive Barnett (2017) and Stephen Collier (2009; 2017) have argued against reductive or ontologizing critiques of government – analyses that, for example, reduce resilience to a single, context-less logic – so too is there a need for affirmative Anthropocene thinking within critical theory to resist this approach. What is needed is not a negation of liberalism per se – a countermovement to or reversal of liberalism that remains caught in what it opposes – but simply the creation of other ways of being in the world. Defining truth and how to live in this context is open to all of the world's masses. It shouldn't be surprising that when ordinary people take these up on their own terms, in their own conditions, what emerges looks far stranger than a design software rendering or academic speculation.

The logical thing should be, not that there's a pluriverse and the only things in it are things I think are good. Rather, the next logical step should be, there is no exterior standard for what a good or bad world is. The task instead is to find out for oneself. Rather than being resilient to or in it, rather than defining oneself in these images, even in the negative ('not resilient'), rather than providing 'solutions' to the Anthropocene, but equally rather than declaring life irreparable, it's possible to simply find new ways of inhabiting the conditions where we are. This entails having the courage and confidence to follow one's own ideas of truth and life, as well as the techniques through which to make them breathe. Such requires thinking a pluriverse free from both liberal modernity as well as any other imposed models for appropriate being.

If the pluriverse is to be a concept relevant for the Anthropocene – in its broad, deep, metaphysical, and societal senses, then it must be radically rethought, expanded, metamorphosed to include the range of worlds and spheres opening up now and which *will* open as diverse people free themselves from liberalism's one world world. These are worlds often in direct conflict with the images said to define pluriversal life. They may express deep ontological and sensible disagreement with ways of life lived by you or I or by leading authors. Just as animism is portrayed as being frightening to the 'modern' mind by Stengers (2018), so too might many more 'modern'-seeming new practices scandalize the liberal mind. Yet they are no less real, no less worthy, than those within the narrow range of what pluriversality is said to encompass.

Such a recognition leads to the question of conflict and coexistence. Rather than interlinking or subsuming all emerging worlds, and rather than imagining peaceful coexistence as the primary relation amongst them – an imaginary common to much pluriverse thinking – what is emerging has other textures, found equally in conflict as well as an even wider gradation of types of relation. Rather than a war between Humans and 'Earthbound' – the former understood as partisans of modern humanist thinking and the latter as diverse in entangled worlds yet homogeneous and willing to work together otherwise (Latour 2017; Viveiros de Castro and Danowski 2018) – conflict in the Anthropocene instead includes possibilities of war *between* worlds, wars over the meaning of what counts as a world, and moreover what can be expected and indeed demanded of such worlds. After all, not every 'world' emerging today sees itself in opposition to so-called Human modernity. Moreover, they may not define themselves in relation – warlike or negating – to liberalism's one world world. Many emerging ways of living simply move on other planes. Likewise, not every world wants to sit at a negotiating table and present itself to a collective. Neither do many worlds seek connection or entanglement with others. Some seek to extract themselves from entanglements which they have experienced throughout their lives as hinderances. Others simply do their own thing. Rather than a bipolar war, then, perhaps the Anthropocene is coterminous with the rise of a multiscalar, social war amongst diverse ways of being in the world and their ideas of truth – as much as the free play between and across such worlds (Tiqqun 2010). From within this broadened definition of world war, the 'front' labeled Earthbound or pluriversal, rather than representing the *whole* of diverse forms of life, now appears as but one front or way of life amongst and emerging alongside countless others. Taking part in such a war entails opening oneself to discover what human life might be like when it finally frees itself of liberal thinking, and forwards its definition of the good life without need for exterior validation or enframing.

Notes

1 In a similar vein Bruno Latour (2017) has tried to think the profusion of multiple worlds as congruent with the Anthropocene. Being 'earthbound,' for him, names a similar vision of a 'multitude of worlds' rather than a single unitary globe.
2 To be clear, I am not speaking of the actual living worlds referenced by many of these authors, but of the latter's characterization of them and their conceptual imaginaries of the pluriverse itself.

References

Adams, R. E. (2014). Natura urbans, natura urbanata: ecological urbanization, circulation and the immunization of nature. *Environment and Planning D: Society and Space* 32(1): 12–29.

Barnett, C. (2017) *The Priority of Injustice: Locating Democracy in Critical Theory*. Athens, GA: University of Georgia Press.

Bennett, J. (2009) Rise of the Preppers: America's new survivalists. *Newsweek*, 29 December. https://www.newsweek.com/rise-preppers-americas-new-survivalists-75537.

Berkes, F. and Folke, C. (eds) (1998) *Linking Social and Ecological Systems: Management Practices and Social Mechanisms for Building Resilience*. Cambridge, UK: Cambridge University Press.

Blaser, M. (2013) Ontological conflicts and the stories of peoples in spite of Europe: towards a conversation on political ontology. *Current Anthropology* 54(5): 547–568.

Bloomberg, M. (2012) December 6. Mayor Bloomberg delivers address on shaping New York City's Future after Hurricane Sandy. Press conference transcript. https://www1.nyc.gov/office-of-the-mayor/news/459-12/mayor-bloomberg-delivers-address-shaping-new-york-city-s-future-after-hurricane-sandy.

Brown, K. (2014) Global environmental change I: a social turn for resilience? *Progress in Human Geography* 38(1): 107–117.

Braun, B. (2014) A new urban dispositif? Governing life in an age of climate change. *Environment and Planning D: Society and Space* 32: 49–64.

Burman, A. (2016) Damnés realities and ontological disobedience: notes on the coloniality of reality in higher education in the Bolivian Andes and beyond. In R. Grosfoguel, R. E. Rosen Velasquez, and R. D. Hernandez (eds), *Decolonizing the Westernized University from Within and Without*. Lanham, MD: Lexington Books.

Burman, A. (2017) The political ontology of climate change: moral meteorology, climate justice, and the coloniality of reality in the Bolivian Andes. *Journal of Political Ecology* 24(1): 921–938.

Burrows R. (2018) Urban futures and the dark enlightenment: a brief guide for the perplexed. In K. Jacobs and J. Malpas (eds), *Towards a Philosophy of the City: Interdisciplinary and Transcultural Perspectives*. London: Rowman and Littlefield.

Chandler, D. and Reid, J. (2018) Contesting the ontopolitics of indigeneity. *The European Legacy* 23(3): 251–268.

Collier, S. (2009) Topologies of power: Foucault's analysis of political government beyond 'governmentality'. *Theory, Culture & Society* 26(6): 78–108.

Collier, S. (2017) Neoliberalism and rule by experts. In V. Higgins and W. Larner (eds), *Assembling Neoliberalism: Expertise, Practices, Subjects*. New York: Palgrave Macmillan.

Crutzen, P. (2002) Geology of mankind. *Nature* 415: 23.

Dalby, S. (2013). Biopolitics and climate security in the Anthropocene. *Geoforum* 49: 184–192.

Dawson, A. (2017). *Extreme Cities: The Perils and Promise of Urban Life in the Age of Climate Change*. New York: Verso.

de la Cadena, M. and Blaser, M. (2018) *A World of Many Worlds*. Durham, NC: Duke University Press.

Descola, P. (2005) *Beyond Nature and Culture*. Chicago: University of Chicago Press.

Edwards, G. A. S. and Bulkeley, H. (2018). Heterotopia and the urban politics of climate change experimentation. *Environment and Planning D: Society and Space* 36(2): 350–369.

Escobar, A. (2008) *Territories of Difference: Place, Movements, Life, Redes*. Durham, NC: Duke University Press.

Escobar, A. (2016) Thinking-feeling with the Earth: territorial struggles and the ontological dimension of the epistemologies of the South. *Revista de Antropología Iberoamericana* 11(1): 11–32.

Escobar, A. (2018) *Designs for the Pluriverse: Radical Interdependence, Autonomy, and the Making of Worlds*. Durham, NC: Duke University Press

Evans, J. (2016). Trials and tribulations: conceptualizing the city through/as urban experimentation. *Geography Compass* 10(10): 429–443.

Evans, B. and Reid, J. (2014) *Resilient Life: The Art of Living Dangerously*. Cambridge, UK: Polity.

Fanon, F. (1961) *Les damnés de la terre*. Paris: Éditions Maspero.

Fath, B. D., Dean, C. A., and Katzmair, H. (2015) Navigating the adaptive cycle: an approach to managing the resilience of social systems. *Ecology and Society* 20(2): 24.

Folke, C. (2006) Resilience: the emergence of a perspective for social–ecological systems analyses. *Global Environmental Change* 16: 253–267.

Foucault, M. (1979) *Discipline and Punish: The Birth of the Prison*. New York: Vintage Books.

Foucault, M. (1984). Polemics, politics, problematizations: an interview with Michel Foucault. In P. Rabinow (ed.), *Essential Works of Foucault* Vol. 1 (111–119). New York: The New Press, 1998.

Fukuyama, F. (1992) *The End of History and the Last Man*. New York: The Free Press.

Gandy, M. (2003) *Concrete and Clay: Reworking Nature in New York City*. Cambridge, MA: MIT Press.

Grusin, R. (ed.) (2017) *Anthropocene Feminism*. Minneapolis: University of Minnesota Press.

Gunderson, L. H. and Holling, C. S. (2002) *Panarchy: Understanding Transformations in Systems of Humans and Nature*. Washington, DC: Island Press.

Hamilton, S. (2017) Securing ourselves from ourselves? The paradox of 'entanglement' in the Anthropocene. *Crime, Law, and Social Change* 68: 579–595.

Haraway, D. (2016) *Staying with the Trouble: Making Kin in the Chthulucene*. Durham, NC: Duke University Press.

Heidegger, M. (2002) Nietzsche's Word: 'God is Dead.' In J. Young and K. Haynes (eds), *Martin Heidegger: Off the Beaten Track*. Cambridge, MA: Cambridge University Press.

Holling, C. S. (2004) From complex regions to complex worlds. *Ecology and Society* 9(1): 11.

Instone, L. (2014) Dialogue. In K. Gibson, D. Bird Rose and R. Fincher (eds), *Manifesto for Living in the Anthropocene*. Brooklyn, NY: Punctum Books, 127–133.

Jacobs, C. K. (2017) Rebuild by design's enduring legacy. *ARCHITECT*, 17 April. http://www.architectmagazine.com/design/rebuild-by-designs-enduring-legacy_o.

Junger, S. (2016) *Tribe: On Homecoming and Belonging*. New York: Hachette Book Group.

Kelly, C. R. (2016) The Man-pocalpyse: Doomsday Preppers and the rituals of apocalyptic manhood. *Text and Performance Quarterly* 36(2–3): 95–114.

Klein, N. (2007) *The Shock Doctrine: The Rise of Disaster Capitalism*. Toronto: A.A. Knopf Canada.

Latour, B. (2017) *Facing Gaia: Eight Lectures on the New Climatic Regime*. Cambridge: Polity Press.

Law, J. (2015) What's wrong with a one-world world? *Distinktion: Journal of Social Theory* 16 (1): 126–139.

Limberg, P. and Barnes, C. (2018) Mimetic tribes and Culture War 2.0. *Medium Intellectual Explorers Club*, September 13. https://medium.com/intellectual-explorers-club/memetic-tribes-and-culture-war-2-0-14705c43f6bb.

Lourie Harrison, A. (2013) *Architectural Theories of the Environment: Posthuman Territory*. New York: Routledge.

Malm, A. and Hornborg, A. (2014) A geology of mankind? A critique of the Anthropocene narrative. *The Anthropocene Review* 1(1): 62–69.

Marres, N. (2018) Why we can't have our facts back. *Engaging Science, Technology, and Society*, 4: 423–443.

McDonnell, T. (2012). Meet the woman tasked with saving New York from the next Sandy. *Citylab*, 10 December.https://www.citylab.com/equity/2012/12/can-woman-save-new-york-next-sandy/4115/

Mignolo, W. D. (2002) The Zapatistas theoretical revolution: its historical, ethical and political consequences. *Review (Fernand Braudel Center)* 25(3): 245–275.

Mignolo, W. D. (2007) Delinking: the rhetoric of modernity, the logic of coloniality and the grammar of de-coloniality, globalization and de-colonial thinking. *Cultural Studies* 21 (2/3): March.

Mignolo, W. D. (2011) *The Darker Side of Western Modernity*. Durham, NC: Duke University Press.

Mignolo, W. D. (2018) On pluriversality and multipolarity. In B. Reiter (ed.), *Constructing the Pluriverse*. Durham, NC: Duke University Press.

Moore, J. M. (ed.) (2015) *Anthropocene or Capitalocene? Nature, History, and the Crisis of Capitalism*. Oakland, CA: PM Press.

Nagle, A. (2017) *Kill All Normies: Online Culture Wars from 4chan and Tumbler to Trump and the Alt-Right*. Winchester, UK; Washington, USA: Zero Books.

Neel, P. (2018) *Hinterland: America's New Landscape of Class and Conflict*. London: Reaktion Books.

Neimark, B., Benjaminsen, T. A., Cavanagh, C., Childs, J., Harcourt, W., Koot, S., Nightingale, A., and Sullivan, S. (2019) Speaking power to 'post-truth': critical political ecology and the new authoritarianism. *Annals of the American Association of Geographers* 109 (2): 613–623.

Nietzsche, F. (1967) *The Will to Power*. New York: Random House.

Nietzsche, F. (2010) *The Anti-Christ*. Auckland, N.Z.: Floating Press.

Ogilvy, J. (2017) The apocalyptic vision of Stephen K. Bannon. *Forbes*, 17 August.https://www.forbes.com/sites/stratfor/2017/08/17/the-apocalyptic-vision-of-stephen-k-bannon/#f02cff346826.

Oslender, U. (2018) Local aquatic epistemologies among black communities on Colombia's Pacific Coast and the pluriverse. In B. Reiter (ed), *Constructing the Pluriverse: The Geopolitics of Knowledge*. Durham, NC: Duke University Press.

Reid, J. (2006). *The Biopolitics of the war on Terror: Life Struggles, Liberal Modernity and the Defence of Logistical Societies*. Manchester: Manchester University Press.

Ronfelt, D. (2007) *In Search Of How Societies Work Tribes – The First and Forever Form*. RAND Corporation Working Paper WR-433-RPC. Santa Monica, CA: Rand Corporation.

Stengers, I. (2018) The challenge of ontological politics. In M. Blaser and M. de la Cadena (eds), *A World of Many Worlds*. Durham, NC: Duke University Press.

Strathern, M. (2018). Opening up relations. In M. Blaser and M. de la Cadena (eds), *A World of Many Worlds*. Durham, NC: Duke University Press.

Sullivan, A. (2019) The abyss of hate versus hate. *New York Magazine*, 25 January.http://nymag.com/intelligencer/2019/01/andrew-sullivan-the-abyss-of-hate-versus-hate.html.

Tiqqun (2010) *Introduction to Civil War*. Los Angeles: Semiotext(e).

Torres, P. (2017) It's the end of the world and we know it: Scientists in many disciplines see apocalypse, soon. *Salon*, 30 April.https://www.salon.com/2017/04/30/its-the-end-of-the-world-and-we-know-it-scientists-in-many-disciplines-see-apocalypse-soon/

Tsing, A. (2015). *The Mushroom at the End of the World: On the Possibility of Life in Capitalist Ruins*. Princeton: Princeton University Press.

Tsing, A. L., Swanson, H. A., Gan, E., and Bubandt, N. (eds). (2017). *Arts of Living on a Damaged Planet: Ghosts and Monsters of the Anthropocene*. Minneapolis: University of Minnesota Press.

Viveiros de Castro, E. (1998) Cosmological deixis and Amerindian perspectivism. *The Journal of the Royal Anthropological Institut* 4(3): 469–488.

Viveiros de Castro, E. and Danowski, D. (2018) Humans and terrans in the Gaia war. In M. de la Cadena and M. Blaser (eds), *A World of Many Worlds*. Durham, NC: Duke University Press.

Wakefield, S. (2014a) Man in the Anthropocene (as portrayed by the film Gravity). *May* 13. https://www.mayrevue.com/en/lhomme-de-lanthropocene-tel-que-depeint-dans-le-film -gravity/.

Wakefield, S. (2014b) The crisis is the age. *Progress in Human Geography* 38(3): 439–456.

Wakefield, S. (2018a) Inhabiting the Anthropocene back loop. *Resilience: International Policies, Practices, and Discourse* 6(2): 77–94.

Wakefield, S. (2018b) Infrastructures of liberal life: from modernity and progress to resilience and ruins. *Geography Compass* 12(7): e12377.

Wakefield, S. (2019) Miami Beach forever? Urbanism in the back loop. *Geoforum.*

Wakefield, S. (forthcoming) *Life in the Back Loop: Experimentation in Unsafe Operating Space.* London: Open Humanities Press.

Wakefield, S. and Braun, B. (2014) Governing the resilient city. *Environment and Planning D: Society and Space* 32(1): 4–11.

Wakefield, S. and Braun, B. (2018) Oystertecture: infrastructure, profanation and the sacred figure of the human. In K. Hetherington (ed.) *Infrastructure, Environment, and Life in the Anthropocene.* Durham, NC: Duke University Press.

Yusoff, K. (2019). *A Billion Black Anthropocene or None.* Minneapolis: University of Minnesota Press.

11

WHAT WOULD YOU DO (AND WHO WOULD YOU KILL) IN ORDER TO SAVE THE WORLD?

Dialectical resilience

Claire Colebrook

Introduction

It might appear, at first glance, that the intensification of post-apocalyptic culture comes as a direct response to the growing awareness of climate change and other twenty-first-century existential threats (such as viral pandemic, resource depletion and runaway artificial intelligence). While theories of extinction and climate change can be traced back to the nineteenth and eighteenth centuries the sense of a crisis of climate change and a predicament of the *species* (and not just the loss of local ecosystems) comes to the fore with the declaration of the Anthropocene. Rather than climate *change* it is now transformation of the earth as a living system that seems to have prompted an intensification of post-apocalyptic culture. Rather than apocalypse, or the sweeping away of this world for the sake of a transformed future, Anthropocene culture can be thought of as *post*-apocalyptic in at least two senses. First, there is no promise of a world after the annihilation of the earth; there is no 'planet B,' and (despite dreams of space migration) we are now united as a single species, not because we share a common human nature, but because we are all in the same condition of facing an end. Second, it appears that we are already living as if the end had already arrived; the question now appears to be how to survive and manage a world that is changed utterly. Approaches to apocalyptic transformation have become managerial. Such exercises in management might be therapeutic (including the growing number of heroic cinematic depictions of saving the world), or they might be strictly techno-managerial, with various practices of re-greening, reverse extinction or geo-engineering promising a sustainable future. It would seem that nothing is more urgent in the current cultural imaginary than *saving* – rather than transforming – the world.

If we accept the premise of the direct relation between an era of mass extinction and 'end of world' culture, then it might also make sense to assume that the problem of resilience becomes urgent and far more focused and attuned to survival at a global level – not just the maintenance of specific systems but saving the world in general. Resilience is no longer a problem of this or that system, or even of this or that world, but takes on global significance as growing awareness of the sixth mass extinction generates a cultural imaginary acutely focused on the possibility of the end of the world. One might imagine that the more we feel threatened the more we focus on shoring up the ruins to become genuinely resilient. We are, it appears, all too aware of our fragility: there are research institutes devoted to saving humanity and the world, prepper subcultures, new genres of cli-fi film and litera-ture, geo-engineering, reverse extinction and projects to re-freeze the arctic. Humanity might appear to be coming to terms with its possible non-existence.

In this chapter I will argue that the contrary is the case: post-apocalyptic culture is a flagrant masking of the thought of extinction, and the dialectic of resilience is anything but an acknowledgment of the fragility of humanity. A long history of thinking about resilience, well before its explicit scholarly articulation, is an expression and cause, and not a response, to a history of extinction. A cultural imaginary of critical fragility – of being poised at the brink of non-being – has enabled and masked what we are now experiencing as the sixth mass extinction. It has been by constantly presenting 'humanity' as both inevitable *and* fragile that the genuine thought of extinction has been warded off. To make this inverse relation more clear, we can think of two quite different examples. The earliest canonical articulations of Western reason in Platonic philosophy defined reason as utterly precarious: sense experience and embodiment draw us to the shadows of the world, and it is therefore the task of reason and philosophy to establish secure foundations that lie beyond the vagaries of mere experience. In this seeming expression of risk and fragility – of reason always being beset by unreason – 'humanity' forges itself as that which is *by right* elevated above a whole series of inhuman risks; reason is always saving itself, both for its own sake and for all those others who have not yet seen the true path of enlightenment (Lycos 1987). Far more recently and flagrantly, White settler culture at one and the same time assumed it was the only true expression of humanity, and yet considered itself to be at risk from miscegenation. White Australia was invaded as *terra nullius,* and yet this supposedly vacant land nevertheless appeared to pose such great risks to white purity that Aboriginal children were stolen from their families and relocated in a mass effort to 'breed out' the color (McGregor 2002). Both reason and whiteness are bound up with a specifically dialectical mode of resilience: both are deemed to be natural and proper – what ought to define humanity in general – at the same time as both are constantly threatened and therefore require strategies of survival. This dialectic is in play both at the level of abstraction, in the Western definition of reason, and in the history of colonization. In both cases the divine right of human reason is assumed to be the natural and proper path of history, even if this history takes the form of an ongoing war on unreason and all its accompanying risks.

Both in its traditional philosophical form, and in the colonialist enterprise, this dialectic of resilience – of constantly having to save human reason because of its right and fragility – is bound up with a quite specific sense of the concept of 'world.' The ideal of reason as a universal logos is bound up with a continuous progression of recognition, ultimately arriving – as we do today – with a sense of a single humanity that occupies a single horizon of sense, all of us affirming each other in differences while aware of a common humanity, and a common predicament. The 'world,' in a strong sense, is an inherently enlightenment notion, bound up with a conception of humanity as properly geared towards unified understanding and a common history. Stephen Pinker, despite his general claims for 'humanism,' and 'enlightenment now,' may be something of an outlier, as his work has received a fair amount of scathing criticism. Even so, his work captures the classic notion that reason justly succeeds despite the odds, and is precisely what is required in an era of fragility:

> The case for Enlightenment Now is not just a matter of debunking fallacies or disseminating data. It may be cast as a stirring narrative, and I hope that people with more artistic flair and rhetorical power than I can tell it better and spread it farther. The story of human progress is truly heroic. It is glorious. It is uplifting. It is even, I daresay, spiritual. It goes something like this.
>
> We are born into a pitiless universe, facing steep odds against life-enabling order and in constant jeopardy of falling apart. We were shaped by a force that is ruthlessly competitive. We are made from crooked timber, vulnerable to illusions, self-centeredness, and at times astounding stupidity ...
>
> As the spiral of recursive improvement gathers momentum, we eke out victories against the forces that grind us down, not least the darker parts of our own nature ...
>
> *(Pinker 2018: 452)*

Pinker's conception of 'recursive improvement' cannot be consigned to the quaint anti-Nietzscheanism of a Harvard psychologist; it aptly captures the twenty-first-century imaginary of the post-apocalypse. There is no world other than this world, and despite its flawed past, it is all we have, and must be saved by way of the very same reason and spirit that appears to have so often gone astray: 'Our ancestors were powerless to stop these lethal menaces, so in that sense technology has not made this a uniquely dangerous era in the history of our species but a uniquely safe one' (Pinker 2018: 295). More specifically, and well beyond Pinker's work, and despite his warning that we should not overly focus on 'existential threats,' one might characterize post-apocalyptic twenty-first century culture as a continuation and intensification of a humanism that is bound up with a modern conception of the world: a single horizon of sense, history and connectedness that looks towards ongoing recognition and legitimation. When we talk today of saving the world, or facing the end of the world, we are not talking about the many worlds that were destroyed in the trajectory that has become known as the Anthropocene, nor are we talking about those modes of existence that are already experiencing the beginning of the end – the worlds of non-Western humans, non-humans and those within the West who have never been able to experience this world as a lifeworld. Rather, the world that offers itself as the unquestioned horizon of resilience, as that which is both worthy of being saved *and* requiring the most urgent attention, is the world of modern global recognition, a world of mass media, humanism and assumed liberal progress. This world is inherently bound up with the dialectic of resilience. Because it is not one world among others but the global project of humanity that ultimately recognizes all others, it is the one world that *must* be saved, even if it is also constantly being threatened with its end.

Both post-apocalyptic narratives and dominant conceptions of resilience that are focused on saving the world occlude the apocalyptic possibility not only of another world, but also of abandoning our fetishized conception of *the world*. Why, we might ask, are depictions of the possible end of the world inevitably located in high profile centers of finance capital, such as New York or London? Why would disaster narratives about the Maldives or the intensely threatened worlds of indigenous peoples not be viable post-apocalyptic fare? Not only are such worlds hardly in the 'too big to fail' league, they are also not *worlds* in the sense that offers a robust demand for resilience. If one thinks about resilience as an attuned responsiveness that constantly negotiates with its milieu in order to sustain itself – neither staying the same nor allowing itself to dissipate – then we have a perfect image of the world in the phenomenological sense. The world is not the planet, and not simply the Western world, but instead a shared horizon of sense: a presupposed unity in which all humans ultimately recognize each other through various cultural expressions. The world is an ideal of liberal rational cosmopolitanism, a moment of historical arrival when every aspect of the globe can be included in one dynamic human endeavor. When the Anthropocene is used in a positive sense – as a moment of historical and geological recognition that will allow us to geo-engineer our way to a sustainable future – the notion of a shared Anthropos is defined by way of resilience: we are bound up with a single world of vulnerability and common purpose. It is this world of resilience that must be saved: this world of inclusion that at once recognizes all others but in doing so becomes every other world's ever so fragile guardian. We must save *the world* precisely because its ability to purvey, include and recognize all worlds exposes its all-encompassing grandeur to the threat of fragmentation. This dialectic is currently playing itself out in the political traumas of 'Brexit' and 'America First.' The ideal of the European union – to maximize trade and minimize conflict – was precisely what made it open to destruction; as systems become all-inclusive they also become subject to increasing pressures. While becoming 'too big to fail' they also become too big to be responsive, allowing any number of marginalized and failing fragments to feel suffocated by the impersonal grandeur of what presented itself as 'humanity.' The grand enlightenment endeavor of rational cosmopolitanism, the liberal ideology of global inclusion and recognition, and the phenomenological conception of a single ideal of legitimation in one shared *lifeworld* – all place so much of a burden on unity that the grand edifice threatens to fall apart. Rather, then, than saying we are living in end times, or that our horizon is simply post-apocalyptic (*after* any sense that there might be a whole new world), it would be more accurate to think of the Anthropocene as a moment of 'peak resilience.' If we arrived at the end of history as a moment when we were no longer torn apart by conflicting ideologies (Fukuyama 1992), we also arrived at the end of the world – where that very unity has precluded any possibility of a simple outside. Alongside a recognition that we are all implicated in the alteration of the earth as a living system and that we have reached a series of 'tipping points,' there is the concomitant assumption that it is precisely *this* damaged world and no other that needs to be saved.

Just as the cinematic experience of flirting with the end of the world only to see it saved reinforces that there can be no other world than this world and that 'we' are too big to fail, so the problem of how much we adapt, adjust and sacrifice in order to remain resilient covers over the thought not only of what might take place after the end of this world, but also what might be possible if one contemplates those forms of existence that are constantly annihilated for the sake of saving the world. One might think of this annihilation on micro and macro levels: the 'world' today relies upon some lives not mattering, as though the value of the world required the non-being of many forms of existence (blackness being the most flagrant example). The campaign that insisted that 'black lives matter' draws upon a long history of theorizing social death; many who exist in this world do not experience it as a horizon of meaning and possibility, but as a system that constantly requires their non-being (Patterson 1982). At a macro level what has come to be known as the Anthropocene, which intensifies the sense of the end of the world and saving the world, has always required the destruction of worlds (especially indigenous worlds). For many, the world ended in 1492 (Lewis and Maslin 2015); the world that is now envisaging (and refusing) its end has long required that many continue to exist in a world already ended, in a life of 'social death.'

Post-apocalypse

We are living in post-apocalyptic times, where the 'post' has several competing senses. The first sense of the post is simple and temporal. There has been apocalypse; after a wave of destruction we now recognize who we are and have been. The Anthropocene is simply the latest version of a thought that goes back at least as far as Theodor Adorno and Max Horkheimer's *Dialectic of Enlightenment* which, before climate change, insisted that looking back on a history of barbarism exposes the damage done to life for the sake of what humanity had thought of as progress. The shudder of the alterity of brute life is covered over by an image of nature as manageable, and then a history of critiques that continue to unmask and refine the representation of the world as ours (Adorno and Horkheimer 2007). For Adorno *negative dialectic* would require thinking about all that has been silenced for the sake of the triumph of reason: art's presentation of harmony would be promissory, indicating something – some resolution – other than this damaged life (Adorno 1973). Today when so much blockbuster cinema depicts humanity triumphing over various annihilation scenarios all the harmony and promise is given with very little sense of that which cannot be rendered human. There is already in Adorno a proto-ecological sense that the world of reason occurs only by silencing life. The robust and unquestioned unity of existence, the increasing rationalization and order of the world, is achieved by a violent negation and denial of anything that would disturb the progress of enlightenment. For Adorno the horrors of the twentieth century are not exceptions but direct consequences of a dialectic that recognizes complexity, volatility and difference only to the point that they can become subject to techno-managerial expertise.

One way, then, to think about *post*-apocalypse is that even before the shock of the Anthropocene the modern capacity to comprehend the disasters that have composed human history required some form of anthropodicy. The trail of wreckage and barbarism that enabled the modern world must come to appear as – despite it all – the best of all possible worlds. This anthropodicy might take the form of Immanuel Kant's notion that despite appearances we must act *as if* the world were on its way to rational cosmopolitanism (Kant 1991) or contemporary notions that climate change is an opportunity – finally – for global justice (Klein 2014). Sometimes this revelatory and transformative sense of the post-apocalyptic is intensified at an imaginary and cinematic level.

A vast number of fictions have depicted a world of resource depletion along with the collapse of complexity, the destruction of mass media and the loss of global political sense. Post-apocalyptic landscapes are littered with dead media, and signs of a lost connectedness. I am Legend (2007) sees Will Smith wandering through a Manhattan where scattered newspapers and flyers remain as the last warning signs of the approaching zombie apocalypse, while Smith returns to his NY apartment to re-play DVDs of a world now lost. The more recent *A Quiet Place* (2018) depicts a remaining family fending for themselves in the countryside, also surrounded by the last fragments and warnings of media culture – posters and newspaper headlines that warn that the monsters operate by sound. It is just this depiction of a world without the comforts of high-consumption capitalism that becomes the occasion for a heroic saving of the world. We are *post*-apocalyptic both because something has been revealed – the destructiveness of human progress – and because that apparent destruction becomes an occasion to save the world. We emerge, rising from the ashes, having cast off whatever it was that was less than human, to discover what we ought to have been all along. The post-apocalyptic is, then, an occasion for intensified resilience; if humans have transformed the planet it is because nature is open to reconfiguration. What has not been considered is that the earth might no longer allow itself to be nothing more than the canvas for human redemption.

After witnessing a history of destruction, or after rehearsing over and over the possibility of destruction (as we do in contemporary cinema), we are faced with questions of resilience: what do we need to do to save a world that is on a path to destruction? The repetition of intensified post-apocalyptic scenarios seems to pose the inevitable question of how, knowing what we know of our damaging past, *we* (the imperiled we) might find our future, or some mode of living on that would require adaptation but not annihilation. How much do we need to give up in order to remain who we are? This is the question of post-apocalyptic culture: after the revelation of the damaging nature of human progress, what might we do to survive? Sometimes this question seems to be answered with utopian felicity: if we abandon capitalism and injustice the true and moral humanity that has always been hampered by some imposed inhuman power might emerge. This is the narrative conceit of The Hunger Games (2012), where an ageing and despotic elite appear as the evil past set against the vibrancy of freedom-seeking youth. In Mad Max: Fury Rd (2015) it is, again, hyper-consuming overlords who are vanquished by a pseudo-indigenous women's collective. Post-apocalyptic dramas all too frequently present an 'end of the world' that is an opportunity for a new world, where the 'new world' is – as always – a revivification of the old. Narrative binaries of old and new, of the 1% versus the 99%, or of capitalism versus 'the people,' allow Anthropos to harbor a redemptive and futural potentiality, where a new world emerges from that fragment of humanity that can forge a brighter future. Some-times it is the very presentation of dystopia that generates its opposite; imagined futures of utter depletion are warning signs of where we might go should we not act now to save the world.

We might think of the post-apocalyptic as a time of realization, where we recognize that there is now a common threat that will bring humanity as a whole together in order that we may all survive. This would be the opposite of a tragedy of the commons – a predicament where everyone's local benefit serves to destroy the conditions of the whole (Gardiner 2011) – and is instead a felicity of the commons. Because all our lives are now so clearly threatened there will be a new solidarity, and hence resilience.

If the first sense to the 'post' is historical and cognitive – a moment of recogni-tion in the wake of the Anthropocene – the 'post' also has a practical and ther-apeutic function, enabling us to continue to be who we are by way of cinematic working through. The 'post-apocalyptic' refers to a specific cinematic and literary genre, perhaps beginning with Mary Shelley's The Last Man (1826) where either the world has ended and humans are left wandering the earth without all that has constituted the complexity of the social fabric of the world, or where such an end is threatened. One might think of the Mad Max series and The Road (both the 2006 novel and the film of 2009) as examples of the former (where humans have been reduced to nomadism, now living without the global interconnectedness of mass media and complex social systems). Threatened end of world films are far more common, where some catastrophe is seen as an invasion or disruption of a highly complex, composed and globally interwoven human world; scenarios of invasion, sudden disaster and destruction allow for narratives of heroic triumph. End of world films are, then, not at all about climate change or extinction, but its oppo-site: human heroic triumph over a foe. Not only do such films allow the times we are living in (of the sixth mass extinction) to be figured as yet one more opportu-nity for humanity to triumph over a threat from outside, they also intensify the assumed absolute value of this world. To present the possible end of the world is to present the demand that it be saved. It is by playing out a series of threatened ends that continued existence becomes ever more urgent. Such films are post-apocalyptic in their refusal of apocalypse; to lose this world is unthinkable. Imagined near-ends are not at all a confrontation with our fragility, but instead forms of doubling down on resilience. Capitalism, imperialism, unbridled consumption and exponential forms of extraction may have brought us to the limit, but it is those same forces in a moralized form that will and *must* save us.

Perhaps the clearest example of this dialectic of resilience would be the forms of post-apocalyptic drama that involve space travel and other similarly heroic efforts to transcend the limits of who we happen to be now for the sake of our proper humanity. In this respect those who wish to think of the Anthropocene as the Capitalocene occupy the same imaginary space as those heroic dramas that see evil corporate powers overthrown for the sake of saving the world. The real and virtual humanity that exists at the margins of capitalism will save the day. In Interstellar (2014) a resource-depleted planet that is subjected to a biopolitical managerialism is surpassed and saved by the new frontier spirit of space travel. In James Cameron's Avatar (2009) a militaristic state focused on extracting 'unobtainium' from Pandora is defeated by a world of nature-attuned quasi-indigenous people who are indeed avatars, models of how we imagine humanity might be. While the manifest fantasy is that of a new world that offers a better model of ecologically sensitive existence, what such narratives reinforce is the prima facie value of *this* humanity, where the new world is (as always) someone else's world *for us*, and where the spirit of triumph and unquestioning self-worth ultimately prevails to save the day. What cannot be considered is the apocalyptic. That this world might not be, and that its end might generate what is not of this world, would be unthinkable.

In this respect it is best to see post-apocalyptic culture as one sustained negation of the sixth mass extinction, and as one ongoing managerial exercise in anthropodicy: what we have been in the past may have been destructive, but future non-being is unthinkable. Even though there is a history of anthropodicy that extends back at least as far as the seventeenth century, the twenty-first century of the Anthropocene intensifies the futural justification of human resilience. *Because* humans as a species have altered the nature of the earth as a living system, it is in their power (and their responsibility) to transform the earth for a better and fully humanized future. The very thought of extinction generates a non-negotiable demand for the future. Humanity may have had its ills but precisely because humanity amounts to *the world* it is too big to fail. The thought experiment of annihilation generates an imperative; it is because we are threatened with not existing that we must continue to be. In a more nuanced version this problem plays itself out in a drama of species bifurcation, where the destructive past of humanity is assigned to an accidental and unfortunate barbarism that reaches fever pitch and that can be overcome by a humanity to come. In films like Elysium (2013) we see the world reduced to indentured labor and controlled by evil corporate powers, with the narrative drive of the film securing the triumph of humanity over enslaving elites. *Avatar,* often cited as a pro-indigenous affirmation of ecological attunement, allows a virtuous and better humanity to emerge (with help from American friends), and in a similar manner the utopian non-violent world of Wakanda in Black Panther (2018) is saved by a white American CIA agent, with the film's hero T'Challa/Black Panther ultimately realizing his true mission of saving modern day America. This is why so many fictions that are ostensibly about the end of the world and climate change are ultimately repetitions of an ongoing human moralism and anthropodicy. Kim Stanley Robinson is perhaps the novelist who has tackled climate change most explicitly in a series of sustained fictions, and yet his narratives are ultimately attempts to forge a point of view that would make sense of the whole, and create a common and just future from the knowledge base that we have spent so many centuries forging. In an interview Robinson makes clear the extent to which science offers a collective promise:

> Our limited personal view of reality means that everyone necessarily has an ideology, which is to say an imaginary relationship to the real situation. You need one, if you didn't have one it would be overwhelming and you'd be incapacitated. So, we have to try to work with what we have. That's where I find science so interesting – it too is an imaginary relationship to the real situation, but it's a group achievement that millions of people have collaborated on over centuries, and its particular imaginary view of the real situation has great power and force, compared to any merely personal opinions. It still has to be expressed in language, sometimes, but it has some powers that are beyond language, and possibly we can put science to use to get a better grip on how we should behave right now.
>
> *(Robinson 2019)*

As with his novels, and despite the science fiction genre with which he works, it is *this world*, a single world – now overtaken by finance – that will redeem itself by way of a revolution driven by increasing knowledge. What such narratives are really about is unquestioned resilience: the very thought of existing otherwise forces an evermore intense insistence on our survival. One might think here of Bladerunner 2049 (2017), where the backdrop of climate change is the canvas upon which a battle between bio-corporations and the miracle of human reproduction is fought. The film is not at all about climate change but instead about the miraculous difference between a life that is not reducible to corporate machination and those near-lives that aspire to be human. In all these cases – as is so typical of the post-apocalyptic – the apocalypse or end is overcome by way of what humanity ought to have been (or *really is*). Climate change becomes the occasion for asserting heroism, frontier spirit and the capacity for humanity, once again, to revive and sustain itself by harnessing the power of others. Capitalism, indentured labor, slavery, barbarism: this will be our future if we do not pursue our better selves. The post-apocalyptic can only think the non-existence of the human as a brief thought experiment for the sake of an evermore resilient living on. By contemplating, ever so briefly, the erasure of what calls itself humanity, the saving of the world becomes ever more urgent.

Dialectical resilience

In addition to drawing a distinction between post-apocalyptic/end of world culture and extinction, it is also important to draw a distinction between two modes of survival or resilience. The first, as I have already suggested, involves sloughing off one's lesser humanity and arriving at a new humanity to come, with the assumption that who one really is – *the real and proper humanity* – has an ultimate right to life, *and* is the privileged being tasked with securing life. The second is almost the opposite, where survival requires destroying oneself *not* to save the world, but because the world is that which has always required one's non-being.

In the first mode of resilience one needs to save the world in order to save one's self, precisely because what it means to be a self is to have a world in the richest sense: a past that gives the present meaning and value, and a future that will unfold the present in its best mode. In the second mode, one must constantly destroy one's self in order not so much to have a world, but because the world is not one's own. As one example of this second mode I want to recall a point that has often been made about the Anthropocene: what appears as *the* end of *the* world for 'humanity' has required the ends of so many worlds. It was by ending the world of many indigenous peoples that 'humanity' (now calling itself Anthropos) trans-formed the earth (Davis and Todd 2017). The condition, today, for thinking about 'a' human species that has progressed to the point of reconfiguring the earth as a living system is the production of a world – a single horizon of sense and a common humanity – that in turn required the annihilation of worlds and peoples:

> one of the deadliest tricks the Anthropocene ever played was making us forget that it was precisely biologically human bodies rendered non-Human that fueled the Human systems which precipitated the Great Acceleration. Climate disaster is the continuing unfolding of the disaster that is the construction of the Human made possible by slavery, colonization, and attempted indigenous genocide.
>
> *(Goldberg 2019)*

This can be thought of in terms of the history of colonization and enslavement, but is also embedded in the daily existence of American life. Here I draw upon the rich tradition of Afro-pessimism that understands blackness not as a skin-deep prejudice that can be overcome by way of inclusion and recognition, but as a negation or social death: there are modes of existence that do not have a world. Calvin Warren has argued that there is no such thing as black being, that the world of being, of things that are available and manageable, is bound up with an imperialism that figured non-being as blackness. This is not only to say that black lives don't matter, or that they are valued less, but that the very conceptions of world and life are composed by negating and destroying black life:

> The Negro is invented, or born into modernity, through an ontometaphysical holocaust that destroys the coordinates of African existence. The Negro is not

a human, since being in not an issue for it, and instead becomes 'available equipment,' as Heidegger would call it, for the purpose of supporting the existential journey of the human being. Black being is the evidence of an ontological murder, or onticide, that is irrecoverable and irremediable.

(Warren 2018: 27)

If one mode of resilience is focused on saving the world in order to save one's self, there is another mode that requires destroying one's self in order to survive, precisely because one has no place in the world. This might appear to be a heavily metaphysical idea but it is made most apparent in literature and film, *and* indicates a mode of existence beyond the metaphysics of the world. In Claudia Rankine's *Citizen: An American Lyric* (2014) the voice that articulates being-in-the-world is a voice of non-being, erasure and constant negotiation with destroying one's self for the sake of living on. To survive is to destroy one's self. In *Citizen* this is depicted in the daily attrition of one's social being, where being a self amounts to destroying one's self.

> You are in the dark, in the car, watching the black-tarred street being swallowed by speed; he tells you his dean is making him hire a person of color when there are so many great writers out there.
>
> You think maybe this is an experiment and you are being tested or retroactively insulted or you have done something that communicates this is an okay conversation to be having.
>
> Why do you feel okay saying this to me? You wish the light would turn red or a police siren would go off so you could slam on the brakes, slam into the car ahead of you, be propelled forward so quickly both your faces would suddenly be exposed to the wind.
>
> As usual you drive straight through the moment with the expected backing off of what was previously said. It is not only that confrontation is headache producing; it is also that you have a destination that doesn't include acting like this moment isn't inhabitable, hasn't happened before, and the before isn't part of the now as the night darkens.

Here, in Rankine's *Citizen*, we see a different sense in which the Anthropocene has demanded the end of the world in order to continue and render itself resilient. Every day, in order for one to survive, one must die a thousand deaths. If the speaker were to turn to her driving companion and say, 'why are you speaking as though I were not here?' or 'why are you speaking as though my blackness does not exist?' the car would screech to a halt, and both bodies would be hurtled forward. Rankine might be read as a lyric poet of 'micro-aggressions,' but these are not the micro-aggressions that can be trained away in company and university training sessions; they are micro in a far more Deleuzian and inhuman sense. Rather than relations of the polity, and rather than the composition of the psyche, micropolitics attends to the forces that compose and decompose the bodies that make up the polity, and the ways in which forces of desire become interests. This would include also the composition of whiteness, of the human, of the world. What is *micro*political is the composition of the world, the ways in which a constituted humanity tethers all value and life to itself. Microaggressions are the ways in which this tethering is repeated and reinforced. Thinking about overcoming this deeply entrenched anti-blackness of humanism requires ending the world. It's not about allowing more freedom, or a future in which 'we' use less and care more. The composition of 'the human' – the global, self-recognizing and reflective humanity – requires an underworld of annihilation, of non-worlds or anti-worlds. There is a world – *the world* – that requires beings who are not granted a world. This may be seen most flagrantly in the history of slavery, but remains in the types of logic articulated by Rankine. If the world were to see and feel this violence would it be able to survive? Is it possible to think of the future not from the point of view of Anthropos and the world, but from a position where 'the human' might need to be destroyed? Writing about black and trans lives mattering, C. Riley Snorton recalls Fanon: 'For some, including and following Fanon, that future effectively means the end of the world. And perhaps black and trans lives' mattering in this way would end the world, but worlds end all the time' (Snorton 2017: 198). Rather than adjusting 'the human' at the edges, and focusing on inclusion, one might pose the question of destruction. How many statues, texts, comportments, habits, institutions, modes of being and senses of human worth would need to be left on the cutting room floor?

The drama of saving the world, the urgency of saving the world, is not only a continuation of Anthropocene imaginaries, it precludes the thought of extinction: not only the sixth mass extinction, but the thousand tiny extinctions that occur every day in so many lives. Extinction, or non-being, is not the same as the end of the world; saving and furthering the world has intensified extinction *and* rendered extinction unthinkable. This is true both on a large geological scale, where transforming the planet to the point of risking the end of the world relied on the annihilation of worlds, and on the scale of daily existence, where the Anthropos of the Anthropocene has been possible only by way of generating lives that cannot matter.

As I have already suggested, one way to make sense of the present is to see the distinction between post-apocalypse and extinction in terms of two conceptions of resilience. The first would be dialectical, where we think about how much we have to abandon or adjust in order to survive, how much we have to adapt in order to avoid the end of the world, or who and what we need to vanquish in order to save the world (with it being self-evident that one would want to save the world). Here, saving the world amounts to saving one's *own* world, or – more specifically – saving the self that is bound up with a sense of *the world*. Here we might think of Martin Heidegger (1995: 196), for whom the animal is 'poor in world.' The animal has some relation to its milieu, but does not have a rich horizon of sense or potentiality. To have a world is not simply to exist and survive, but to have a present where one's possibilities are inflected by a past and driven by an anticipated future. The end of the world would amount to bodies wandering the earth, merely existing and surviving without any care or conception of the temporality – the rich past – through which earth becomes world. One might reflect on how many post-apocalyptic landscapes are the end of the world insofar as only humans remain. If there are merely bodies then we have reached the end of the world: if we cannot recognize those things and texts that composed a world of sense then we might as well be animals. (Films such as Planet of the Apes [1968], The Road [2009], Oblivion [2013], and Blade Runner 2049 [2017] all rely on the heavily burdened meaning of things: what must *not* happen is that the things that compose our world become nothing more than mere matter. The Statue of Liberty must not be detritus washed up on a beach; someone must be left to remember a world of books.) Our world is bound up with the texts and things that allow us to see the world as having been there for others in the past, and as an archive for those who will be in the future; a world without that archive would be the end of the world. This mode of resilience thinking would require us to ask: how much of this archive do we need to keep being ourselves?

Non-dialectical resilience

The second way of thinking about resilience would be destructive; what might it be like to think in terms of those whose existence already requires an ongoing annihilation of one's self? In a book on what we owe to animals, Christine Korsgaard asks us to imagine a species of greater intellect invading our world and taking our lives simply on the basis of their superiority; we would still feel our life and world valued *to us* (Korsgaard 2018). Her broader philosophical point is that from the point of view of those whose lives we take and consume, our lives do *not* have greater value. It makes sense for us to care more about our world than we do the worlds of others, but that is not an absolute so much as a tethered value. Korsgaard's point concerns the relation between humans and animals, and she does explore some truly important differences, including the human capacity to reflect upon one's choices. To ask whether this capacity for reflection makes our lives more important is, she insists, nonsensical. The capacities we attach to living well are important for us, and who counts as 'us' is neither self-evident nor stable. The 'we' can refer to 'we' humans as a species, but it could also refer to 'we' who inhabit the planet – including all life forms. What lies outside the range of Korsgaard's argument is the intra-human question of the 'we' or 'us.' The capacities we value make our lives important to us, but are there capacities and modes of existence that not only are *not* important for others, but that also require the *non-being* of others? Central to Korsgaard's broader project are the human capacities for reflecting upon one's choices, and having a strong sense of personhood. To what extent has the history of personhood – the grand enlightenment project of universal humanism – required the annihilation of lives deemed to have no moral standing, and to what extent have forms of existence demanded of 'fellow creatures' precluded moral standing? It may be argued that while the history of enlightenment has been tied to slavery and the annihilation of many lives deemed to be non-human, this need not be the case. Or, even if the grand march of civilization required some degree of barbarism to achieve the technological sophistication we have today, we could argue that the Anthropocene offers the possibility of a new humanity. We are now bound to this earth, and can no longer ignore our common planetary predicament.

I want to suggest, though, that in the current crisis of the Anthropocene saving who we are is essentially bound up with the destruction of others – both human and non-human. What would it require to save *who we are,* beings whose sense of choice, self-formation and autonomy has – for so long – required increasing technological maturity and conditions of privacy that would maximize the sense of being a law unto oneself? If we now know the cost of the tradition of Western individualism, which relied upon slavery, colonization and the consumption and the transformation of the planet that is now known as the Anthropocene, can we maintain that cost for 'our' own future? This question of sustainability requires reconfiguring the way we think about resilience. If we took very seriously what it might entail to save the world, then we arrive at two conflicting senses of the world. Reducing the impact that humans have had on the planet might begin with driving more efficient vehicles, eliminating plastics, going vegan and moving towards renewable energy, but ultimately the world that we currently live in would probably have to end in order to save most of the living systems that compose the earth. One might say then that one would need to destroy one's own world to save the world. At what threshold does saving what has come to be known as the world allow for (and demand) one's end? What would you do (and who would you kill) to save the world? What has come to be known as 'the world' has already required so much death, loss, annihilation and erasure. Is there a point of 'peak consumption,' where there is no longer enough of a milieu to sustain the world in anything like its current form?

The world of resilience – the world that constantly manages its milieu in order to save itself – is always already bound up with the end of the world. In order to arrive at the Anthropocenic moment, where life as a whole now requires some degree of global rescue, there must already have been the widespread destruction of worlds. The world that is now obsessed with averting the end of the world has always been world-ending, always demanded the desolation of other worlds. One might therefore draw a contrast between a moment of peak resilience, where 'humanity,' confronts the question of how much it must sacrifice in order to save its world, with a counter-history of those who have had to exist without world in order to live, or those whose ongoing non-being underpins the world. We might draw a contrast between 'man' who must now confront what must be given up to save the world, and those who have already died a thousand tiny deaths in order to exist in a world that constantly demands their end. Those who are dead within this world – what Orlando Patterson (1982) refers to as 'social death' – provide another way of thinking about ending the world: perhaps for some bodies and modes of existence what has come to be known as 'the world' has already required one's annihilation. This insight is at the heart of a series of Afro-pessimist texts that have embraced the possibility of the end of the world. At the very least one might argue that the world as it is currently composed – having emerged from a long tradition of anti-blackness that included slavery, colonization and their aftermath – that the world needs to end in order for many lives to become viable or to matter. Here, demanding that black lives matter would require not simply inclusion in the constituted polity, or a recognition that black lives have the same worth, but a polity no longer built on worth. The very possibility of disposable life, of non-life and non-being, is intimately entwined with a history of anti-blackness whereby the world becomes a horizon of ever-expanding sense, purposiveness and techno-managerial efficiency. The Heideggerian conception of world, where I am nothing more than an open and decisive unfolding of projects, and where others are also bound up in my common horizon of sense, required the disdain and erasure of lives that were disparate, dispersed, not oriented to projects of self-definition. Writing about Mayotte Capecia's *Je Suis Martiniquaise,* Franz Fanon aptly captured this double sense of world, where a certain conception of blackness is tied to being without world, while whiteness definitively *has* world where things have varying degrees of value:

> I am white: that is to say that I possess beauty and virtue, which have never been black. I am the color of the daylight....
>
> I am black: I am the incarnation of a complete fusion with the world, an intuitive understanding of the earth, an abandonment of my ego in the heart of the cosmos, and no white man, no matter how intelligent he may be, can ever understand Louis Armstrong and the music of the Congo. If I am black, it is not the result of a curse, but it is because, having offered my skin, I have been able to absorb all the cosmic *effluvia.* I am truly a ray of sunlight under the earth ...
>
> And there one lies body to body with one's blackness or one's whiteness, in full narcissistic cry, each sealed into his own peculiarity – with, it is true, now and then a flash or so, but these are threatened at their source.
>
> *(Fanon 2004: 31)*

In addition to thinking that one must end *this world* of anti-blackness to make way for a world in which resilience is no longer achieved by constantly annihilating all those perceived inhumanities that threaten the supremacy of one's being, one might also consider forms of existence that have up until now required one's non-being for the sake of the world, and that may well flourish in the utter absence of world.

Here, I want to look at two texts that give an account of killing one's self for the sake of saving the world, along with *ending the world and killing one's self in order to live.* The first text is Jordan Peele's recent *Us* (2019), which continues the thread of his earlier *Get Out* (2017). In *Get Out* an ageing white elite continues to live by transplanting their brains into vibrant black bodies. The hijacked bodies never quite allow themselves to be fully taken over, and can shed a tear for being who they now are; their subjectivity exists only in a visceral rebellion against the white interiority they now harbor. This is an odd repetition of William Blake's depiction of the piety of Christian humanism: 'I am black, but O! my soul is white.' White interiority is a form of being destroyed while continuing to exist, in a world that is not one's own. One poignant scene depicts a servant shedding a tear, her body desperately at odds with the white brain that has been transplanted into her skull. At one allegorical level the film depicts a condition so often described by Fanon, where it is white anti-black subjectivity that has overtaken black bodies. But there is also a less allegorical dimension: the young black photographer whose body is being purveyed and set as a prize at a garden party comes to recognize that there are no bodies around him with whom he shares a world. In order to 'get out' the central character has to kill everyone around him, including 'grandma' who has conveniently overtaken the body of the black housemaid Georgina. It appears you have to destroy those around you to get out, and that the entire seemingly pro-black edifice is there to sustain white interiority and vision. (The central body-swap will be between a blind white photographer and the young, gifted and black central character Chris.) The allegorical dimension – whiteness installed in black bodies – is troubled by an almost literal dimension: you must not be yourself, you will be trapped in yourself. Living on entails not being who you are.

In the later *Us*, surviving requires killing yourself, both – initially – the self who is your horrific doppelgänger, but ultimately also killing the real you, the you that was dragged down and chained to an underworld. *Get Out* and *Us* take part in a long literary tradition that runs counter to the notion that being a self is bound up with being-in-the-world. Instead, in order for the world to continue one must constantly annihilate oneself. In *Us* a successful black family becomes the victim of a home invasion, where their double – a demonic version of the mother, father and two children – need to be vanquished in order for the family to live on. The central character, the mother, is handcuffed by her double, and told that an entire species of doubles had been created in order to control the world. Unfortunately, that plan didn't appear to work and the doubles now live in an underworld, surviving on nothing more than another species – rabbits – who live and multiply along with their tethered humans. The first allegorical dimension is apparent; there's another humanity, negated, *held in the underworld,* now abandoned, buried, forgotten but nevertheless still tethered to the world. Created to manage the world, they now only exist as an aftermath and inadmissible double. Their only exit is through a violent cutting; they make their way up to the world with scissors, dressed in red suits that render them into large scale versions of the human figures in the 'hands across America' graphic of 1986. They are the ghastly doubles of a redemptive humanism, a humanism that wants to see everyone contained in a home of their own. They must take on the figural form of humanist tradition, a figure of colorless hand-holding human solidarity. A second and more profound allegorical dimension unfolds where the survival of the middle-class black family requires *killing one's self and one's own kind.* At the point at which the mother is faced with killing the doubles of her own children there is a hesitation, until her son takes over the role. Those two allegorical layers make clear that what is given as the world that must be saved requires and has required burying, killing and tethering a race of inhuman humans, with no voice and no life other than that of the ghastly doubling of the world.

The final allegorical dimension is where the seeming humanism of the first two registers falls apart. In a film that is about killing yourself in order to save the world, one might say that all post-apocalyptic films are about destroying or sacrificing that aspect of our being that threatens the future. In a last-minute plot twist it turns out that the central character who encountered her underworld double as a child was dragged into captivity while the speechless doppelgänger assumed her role in first world life. Gradually gaining powers of speech, she also becomes fully human; it is, then, the doppelgänger who must kill her first self in order to save herself, her family and the world. At a glib level one might see this as the reality of black life; *do not exist, do not matter, or you may not survive.* A possible morality might emerge whereby one might demand that black lives matter, and their non-being should not be the condition for the coherence of the world. Profound as that may be, there is something profoundly metaphysical at work in the implications of social death, where blackness signals a negativity of utter extinction that is not that of the end of the world but is absolutely worldless – without speech and without futurity. The self that has been tethered, held, forgotten and buried turns out not to be your monstrous other but you yourself; you have to kill yourself to save the world. This is, as the first two layers of allegory make clear, the condition of black life. Rather, though, than bringing that life into the world, humanizing and untethering its abandonment, we might see another mode of resilience. It may be necessary to contemplate your extinction for the sake of a world you thought was your own.

My second example of a text that recognizes the ways in which selves have been forged in conditions of non-being is Toni Morrison's (2015) *God Help the Child.* The novel begins with the narrator – who is proud of having been able to 'pass' – expressing her utter horror at the darkness of her newborn child's complexion. Despite her history and efforts, her own blackness has returned to her in the form of a child she can barely own. The rhetoric that composes her world requires a negation of her own being. She speaks as if she were tethered to a world that cannot admit her existence. Here, as elsewhere in Morrison's fiction, survival in *this* world amounts to destroying or negating one's own being. There's nothing *resilient* or *managerial* here; it is only by way of *not being* that one might survive. One might say, following a rich tradition of Afro-pessimism (and despite Morrison's own avowed humanism where we all belong to the one race of humankind [Morrison 2017]) that the being of blackness is one of social death. Even outside of *Beloved,* where all one can do for the care of one's kind is to secure their death, Morrison's world is one in which one must kill oneself in order to survive. The narrator who opens and concludes the novel can only see giving birth as an event of disenchantment: 'Listen to me. You are about to find out what it takes, how the world is, how it works and how it changes when you are a parent' (Morrison 2017: 253). Her daughter, who managed to flourish until she was severely beaten by a woman she had falsely accused of sexual abuse as a child, looks at parenting as the possibility of a new world: 'A child. New life. Immune to evil or illness, protected from kidnap, beatings, rape, racism, insult, hurt, self-loathing, abandonment. Error-free. All goodness. Minus wrath. So they believe.'

It is just that hope of another world that is radically undermined by both the narrator, and a prior narrative of self loss. After being abandoned by her lover Bride starts to disappear – her breasts and the pierced holes in her body are suddenly gone. The narrative ends with her mother predicting the same dull round of dispossession.

Morrison describes a world in which both mother and daughter have to deny or lose their being in order to exist, in which the hope one has for one's child is that there will be another world, *not* this world, or that one might not be recognized as black. One's existence *as who one is* becomes non-negotiable, impossible, destructive. One might note that in a literary and philosophical tradition quite different from end of world and post-apocalyptic epics, Morrison's fiction detailed a mode of being where one needed to destroy or negate oneself in order to survive. *God Help the Child*'s first paragraph is spoken by a mother desperately trying to assuage the guilt she feels at having not been able to touch or own her daughter:

> It's not my fault. So you can't blame me. I didn't do it and have no idea how it happened. It didn't take more than an hour after they pulled her out from between my legs to realize something was wrong. Really wrong. She was so black she scared me. Midnight black, Sudanese black. I'm light-skinned, with good hair, what we call high yellow, and so is Lula Ann's father. Ain't nobody in my family anywhere near that color. Tar is the closest I can think of yet her hair don't go with the skin. It's different – straight but curly like those naked tribes in Australia. You might think she's a throwback, but throwback to what?
>
> *(Morrison 2015: 1)*

There are, then, two ways of thinking about killing yourself to save the world. The Anthropocene has intensified the dialectical mode of resilience of 'Anthropos' where one must make sacrifices in order to live; 'humanity' would mark and destroy as 'inhuman,' so much of the world it sought to save and manage. In quite another mode, killing yourself to save the world might be seen in terms of blackness, in a world composed by way of anti-blackness: the world demands that one *not be*. 'Anthropos' continues to save the world by demanding the non-being of others. Blackness makes its way in the world only by not being.

If 'anthropos' continues to save the world, he does so by means of the destruction of worlds. This destruction is a necessary component of a world of dialectical resilience, where black bodies were taken up as so many things and resources that the world needed to manage in order to survive. Rather than negotiating how much of ourselves 'we' need to sacrifice in order to save the world (which is the dialectical form of resilience that is intensified in the Anthropocene) it might be worth adopting the point of view of all that has had to constantly destroy itself in order to occupy the world. One might think of this other mode of resilience, *not* as one of negotiating just how much of our world we can continue to destroy in order to live on and save ourselves, but recognize that 'the world' has already demanded the end for so many modes of existence to the point where the end of the world might be the opening to something other than constant annihilation. Rather than having to destroy oneself constantly in order to be in the world, there might be a mode of existence *without world*. Accepting that there is no single horizon of sense, no assumed common felicity, but always an exposure to what may not further one's own existence, there may well be a form of life and resilience after the end of the world.

This is what I take Afro-pessimism's conception of social death to be, an awareness not so much that one does not have a world or belong in the world, but that the world demands one's non-being. Currently this form of existence is utterly tragic, constantly resulting tracing the wake of black lives *not* mattering. Even so, Afro-pessimism also offers a positive sense of the end of the world, where non-being and worldlessness provoke thought to move beyond the world. This is also how I understand Calvin Warren's (2018) *Ontological Terror*: blackness marks an unthinkable non-being that cannot be held within humanism. Instead, such blackness takes hold and – if taken up – might require the end of the world. As I have already suggested, Afro-pessimism may be a late twentieth-century and early twenty-first-century philosophical movement, but the articulation of blackness without world has a much longer history that offers a counter-history to an enlightenment narrative of increasing resilience. From slave narratives to their aftermath in the texts they inspire today, there is a space apart from the world, an experience of what it means to be held without the assumption of a common humanity.

In the history of enlightenment, humanity has managed to generate a world – constantly saving itself and its milieu by way of the destruction of worlds and others. What this has established is a resilience of 'too big to fail.' We allow fragments of the world to disappear, but we do so for the sake of who we are. There is a constant negotiation of dialectical resilience: how much do we need to sacrifice in order to sustain who we are? This can range from the removal of monuments, the abandonment of plastics and fossil fuels, but also – as has always been the case – to sacrificing what is never regarded as fully human for the sake of the human. This would yield the post-colonial negotiations of the present: how much do 'we' the first world need to abandon in order to live, and how much do we allow those in lesser worlds to take on the technology that at once allowed the world to seem to be a global interconnected unity, while allowing worlds to become extinct? We could keep playing the post-apocalyptic game of imagining that all we need to do is use less, and include others, and our proper humanity will save the future. Alternatively, we could work in the opposite direction. If the worlds 'we' destroyed could speak, what might they allow to live? I suspect the answer would not bode well for 'us.' If, faced with the humanity we tethered and held, would the anthropos of the Anthropocene appear to have any right to life? Rather than surviving by sloughing off what we would like to think of as accidental and monstrous, it would be better to acknowledge the monstrosity of a humanity that has demanded extinction for the sake of the world.

References

Adorno, T. (1973) *Negative Dialectics*. London: Continuum.

Adorno, T. and Horkheimer, M. (2007) *Dialectic of Enlightenment: Philosophical Fragments.* Stanford: Stanford University Press.

Davis, H. and Todd, Z. (2017) On the importance of a date, or, decolonizing the Anthropocene. *Acme: An International Journal for Critical Geographies* 16(4): 761–780.

Fanon, F. (2004) *The Wretched of the Earth*. New York: Grove Press.

Fukuyama, F. (1992) *The End of History and the Last Man*. New York: Free Press.

Gardiner, S. M. (2011) *A Perfect Moral Storm: The Ethical Tragedy of Climate Change*. Oxford: Oxford University Press.

Goldberg, J. (2019) Ethics and troubled kinship in the wake of still-unfolding disaster. *ASAP/Journal*. May 2. http://asapjournal.com/ethics-and-troubled-kinship-in-the-wake-of-still-unfolding-disaster-jesse-a-goldberg/Heidegger, M. (1995) *The Fundamental Concepts of Metaphysics: World, Finitude, Solitude*. Bloomington: Indiana University Press.

Kant, I. (1991) Idea for a universal history with a cosmopolitan purpose. In I. Kant, *Political Writings*. Cambridge: Cambridge University Press, 41–53.

Klein, N. (2014) *This Changes Everything: Capitalism Vs the Climate*. New York: Simon Schuster.

Korsgaard, C. (2018) *Fellow Creatures: Our Obligations to the Other Animals*. Oxford: Oxford University Press.

Lewis, S. L. and Maslin, M. A. (2015) Defining the Anthropocene. *Nature* 519: 171–180.

Lycos, K. (1987) *Plato on Justice and Power: Reading Book I of Plato's Republic*. Albany: SUNY Press.

McGregor, R. (2002) 'Breed out the colour' or the importance of being white. *Australian Historical Studies*, 33(120): 286–302. Morrison, T. (2015) *God Help the Child*. New York: Vintage.

Morrison, T. (2017) *The Origin of Others*. Cambridge: Harvard University Press.

Patterson, O. (1982) *Slavery and Social Death: A Comparative Study*. Cambridge: Harvard.

Pinker, S. (2018) *Enlightenment Now: The Case for Reason, Science, Humanism, and Progress*. New York: Viking.

Rankine, C. (2014) *Citizen: An American Lyric*. Minneapolis: Graywolf Press.

Robinson, K. S. (2019) Interview: Kim Stanley Robinson's lunar revolution: the sci-fi author on geopolitics and his latest novel, 'Red Moon.' With Eliot Peper. *Chicago Review of Books*, January 7. https://chireviewofbooks.com/2019/01/07/kim-stanley-robinsons-lunar-revolution/ Accessed June 3, 2019.

Snorton, C. R. (2017) *Black on Both Sides: A Racial History of Trans Identity*. Minneapolis: University of Minnesota Press. Warren, C. L. (2018) *Ontological Terror: Blackness, Nihilism, and Emancipation*. Durham: Duke University Press.

Motion pictures

A Quiet Place (2018) Bay, M., Form, A. & Fuller, B. (Producers) & Krasinski, J. (Director). United States: Paramount Pictures.

Avatar (2009) Cameron, J. & Landau, J. (Producers) & Cameron, J. (Director). United States: 20th Century Fox.

Black Panther (2018) Feige, K. (Producer) & Coogler, R. (Director). United States: Walt Disney Studios Motion Pictures.

Bladerunner 2049 (2017) Kosove, A., Johnson, B., Yorkin, B. & Yorkin, C. S. (Producers) & Villeneueve, D. (Director). United States: Warner Bros. Pictures.

Elysium (2013) Block, B., Blomkamp, N. & Kinberg, S. (Producers) & Blomkamp, N. (Director). United States: Sony Pictures Releasing.

Get Out (2019) Pelle, J. (Director) & McKittrick, S., Blum, J., HammJr., E. H. & Peele, J. (Producers). United States: Blumhouse Productions, QC Entertainment, Monkeypaw Productions.

I Am Legend (2007) Goldsman, A., Lassiter, J., Heyman, D. & Moritz, N. H. (Producers) & Lawrence, F. (Director). United States: Warner Bros. Pictures.

Interstellar (2014) Thomas, E., Nolan, C. & Obst, L. (Producers) & Nolan, C. (Director). United States & United Kingdom: Paramount Pictures & Warner Bros. Pictures.

Mad Max: Fury Road (2015) Mitchell, D., Miller, G. & Voeten, P. J. (Producers) & Miller, G. (Director). Australia & United States: Warner Bros. Pictures.

Oblivion (2013) Chernin, P., Clark, D., Henderson, D., Kosinski, J. & Levine, B. (Producers) & Kosinski, J. (Director). United States: Universal Pictures.

Planet of the Apes (1968) Jacobs, A. P. (Producer) & Schaffner, F. J. (Director). United States: 20th Century Fox.

The Hunger Games (2012) Jacobson, N. & Kilik, J. (Producers) & Ross, G. (Director). United States: Lionsgate Films.

The Road (2009) Wechsler, N., Schwartz, S. & Schwartz, P. M. (Producers) & Hillcoat, J. (Director). United States: Dimension Films.

Us (2019) Pelle, J. (Director) & Blum, J., Cooper, I., McKittrick, S. & Pelle, J. (Producers). United States: Monkeypaw Productions, Perfect World Pictures.

INDEX

Printed in the United States
by Baker & Taylor Publisher Services